FIX YOUR BIKE

REPAIRS AND MAINTENANCE FOR HAPPY CYCLING

JANE MOSELEY AND JACKIE STRACHAN

PORTICO

CONTENTS

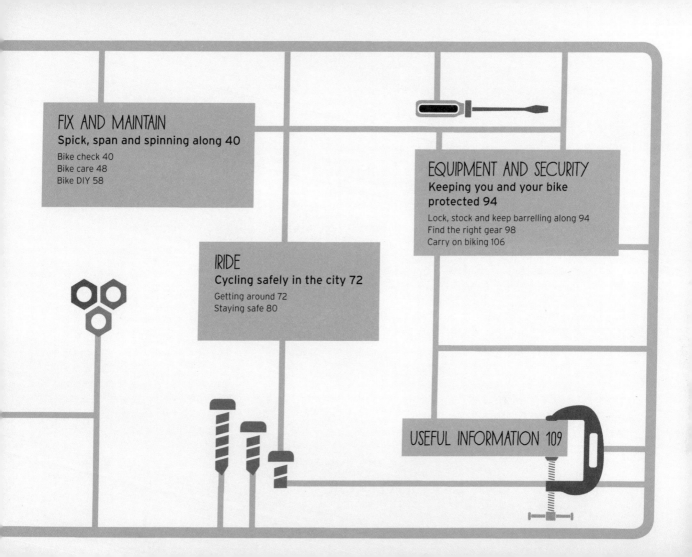

THE WHEEL BENEFITS OF CYCLING

More and more people are choosing two wheels over four to get around town, both for leisure and work purposes. Urban cycling ticks all the boxes, just see what it can do for you.

It's good for you from a freedom, fiscal, fresh air and physical fitness point of view: your bank balance, metabolic rate, endorphin levels and immune system will all go up. You will get more sleep at night after all the exercise, you will reduce the stress you put on your knees if you have been jogging to work, and research has shown that your mental fitness is also likely to increase. It's a win/win and it's also a win/lose, as you'll be burning not money or fuel but calories - up to 600 an hour, depending on your size, speed and steepness of incline. You will save on commuter emissions, personal costs, stress and time (in most places), and what's more (or less - OK, that's enough) you won't be adding to the planet's sky-rocketing pollution levels. Your upper body tone and leg muscles will firm up and you will get fitter as your wallet gets fatter. You'll be leaner and the world will be greener. It's fun and free. Result.

There's no need to put yourself and your bike under excessive (tyre) pressure and start by cycling every day. A few times a week is fine (once a week at first) and you'll soon get the bug, and not the one from coughing commuters. And if you don't carry a lock on your regular commute, and leave it at work instead, you'll resist the temptation to hit the bar on the way home because you can't leave your bike unlocked. Like a Tantalus but in reverse. Once safely parked and locked up (your bike, not you), you might just head to the gym instead.

Cycling is a low-impact, high-reward form of travel. See how it works out as you work out on www.cycletoworkcalculator.com

GREAT BIKING MOMENTS

1696

1696
French mathematician Jacques Ozanam throws down gauntlet to inventors to come up with a horseless carriage in his book *Recreations Mathematiques et Physiques*.

1817
Eccentric German baron Karl von Drais patents his steerable, two-wheeled *Laufmaschine* (running machine), aka *draisine* or hobby/dandy-horse.

C.1840
Englishman William Sawyer invents a four-wheeled velocipede using foot treadles. Around the same time, Scotsman Kirkpatrick Macmillan designs what is credited (by some) to be the first pedal cycle.

EARLY 1860S
Pierre Michaux (or son Ernest) adapts a draisine with cranks and pedals. Pierre Lallement also credited with a similar invention – the first 'boneshakers'.

1868
Michaux forms a company with the Olivier brothers to mass-produce two-wheeled velocipedes.

1817   **c1840**

1869
Jules Suriray patents ball bearings for wheel hubs.

1870S
The unfeasibly large high wheel bicycle (aka penny farthing or 'ordinary') is developed.

1874
James Starley designs the tangent-spoke wheel, used by builders of almost all bikes ever since (and borrowed by the automobile, motorbike and aviation industries).

1860s

1868

1869

1870s

1874

1878

1885

1888

1890s

1896

1903

1905/6

1962

1970s

1878
William Grout invents the first folding bike (possibly).

1885
John Kemp Starley launches his Rover Safety Bicycle with a chain-driven rear wheel and two similar-sized wheels. His company will become the British Rover automobile company. The increased safety is matched by a rise in popularity.

1888
Scotsman John Boyd Dunlop invents the pneumatic tyre, and the bicycle is now on a roll.

1890S
Lightweight bikes with diamond-pattern frames, chain drives, pneumatic tyres and brakes are now being mass-produced.

1896
William Reilly develops a two-speed internal gear hub.

1903
The first Tour de France is held.

1905/6
Paul de Vivie invents the first derailleur gears.

1962
Alex Moulton releases his first iconic small-wheeled bicycle; it becomes a Swinging Sixties trendsetter. Al Fritz designs the Schwinn Sting-Ray, a wheelie bike that resembles a chopper motorbike.

1970S
Mountain bikes are developed in California; in 1977 Joe Breeze produces his first Breezer. Children riding wheelie bikes and racing on dirt tracks inspire the development of BMX.

CHAPTER ONE

GET IN GEAR
ALL YOU NEED TO BEGIN YOUR JOURNEY ON TWO WHEELS

GET A HANDLE ON THE BIKE FOR YOU

So you're buying a bike? That's great. You might be looking to add a reliable town bike to an already extensive bike wardrobe, or perhaps you haven't been on a bike since you were six. Now that cycling has expanded into different areas of specialization, there is now a wide range of different types of bike available. But before you go shopping, here are a few things to think about.

KNOW YOUR BIKE

Before you start looking into the different kinds of bike available, here's a brief guide to the major components. You'll notice that for every pro there seems to be a con, and vice versa, but hopefully this review will help you to home in on the issues that are key for you. (For buying a bike the right size, see page 32.)

BIKE BASICS

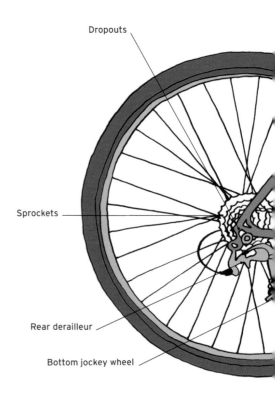

Dropouts

Sprockets

Rear derailleur

Bottom jockey wheel

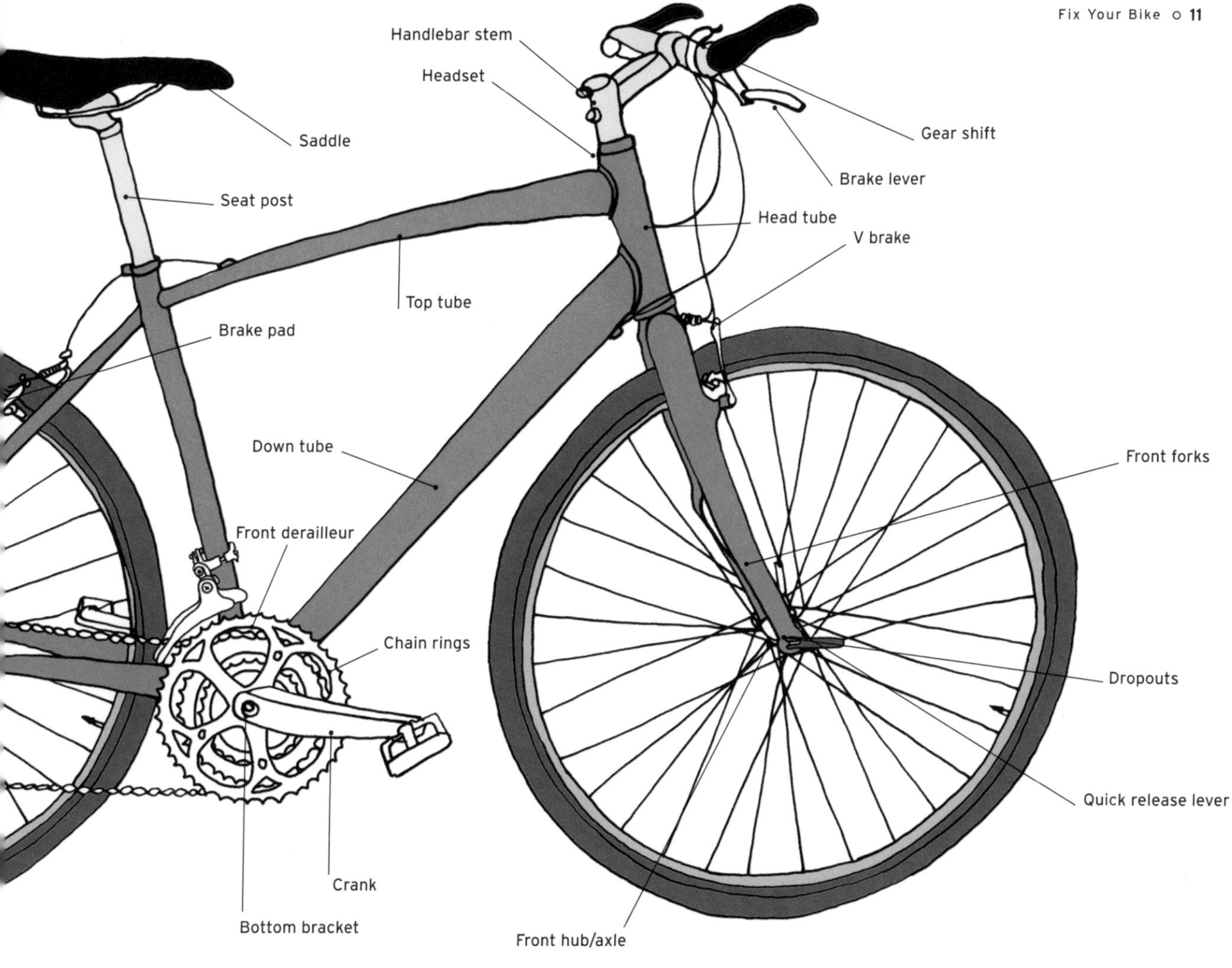

Handlebar stem

Headset

Saddle

Gear shift

Brake lever

Seat post

Head tube

V brake

Top tube

Brake pad

Down tube

Front forks

Front derailleur

Chain rings

Dropouts

Quick release lever

Crank

Bottom bracket

Front hub/axle

THE RIGHT BIKE FOR YOU

1 Where will you keep it? If you need to climb stairs to store it - on a landing for example - weight immediately becomes an issue. And space? If it's at a premium, then a folding bike may be the answer.

2 If you are commuting, what is your route like? Hilly? Will you have to carry your bike up or down any steps? You'll need a reasonable range of gears and again won't want a bike that's too heavy.

3 You may have some Lycra lurking at the back of the wardrobe, but the chances are that you will be cycling in 'normal' clothes a lot of the time. If that means sometimes in a dress or skirt, a step-through frame might be worth thinking about.

4 How much stuff will you need to carry? A backpack is useful, but can make your back hot and sweaty and drag on the shoulders. So how about panniers or maybe a basket?

5 And finally, how much do you want to spend? Bikes are available at all prices and second-hand (from a reputable source) is also an option. Will you have to leave your bike chained up unattended for hours at a time? If you are worried about returning to find just the front wheel, all forlorn but still securely locked to its post, maybe now's not the time to splash the cash.

Do you get what you pay for? On the whole, yes, although you may pay extra for something fashionable in cool colours. It's probably advisable to steer clear of the absolute budget end of the market and get a robust bike that will last. Ask people about their bikes, check out your local bike shop and when you're ready to buy, perhaps take a cyclist friend along with you.

FRAME

The basic diamond-pattern frame shape has been around for over 100 years, although there have been plenty of other variations since then. The top tube was traditionally horizontal with the road on men's bikes, and dropped for women, so they wouldn't have to do anything as unladylike as swing a leg over to mount the bike – rather difficult clad in an ankle-length skirt. These days, the degree of slope of the top tube is often more a function of the design rather than a hint about which gender should ride it. Bikes are made specifically to suit women's physical proportions as well as men's and with step-through frames, but there is no reason why a woman should not buy man's bike if it fits her and suits the kind of cycling she wants to do. And a man with a bad back might find that a step-through frame is just what the doctor ordered.

The weight of the frame affects the weight of the bike quite considerably and depends on what it is made of.

Aluminium alloy is light and won't rust, but is less durable than steel; you will also feel bumps and lumps more when you ride over them than with steel. It is the cheapest of the lighter materials and is now commonly used for frames.

Steel is strong, but can rust and can be heavy on bikes at the cheaper end of the market. Nearly all frames used to be made of steel, but far fewer are these days.

Carbon fibre is light and strong but expensive, and can crack if knocked hard or when involved in a crash.

Titanium is light, strong and won't rust, but is very expensive, so is used for high-end bikes only.

Good to know
The brand name on the frame is usually that of the company that assembles the bike, having bought in the components from other companies.

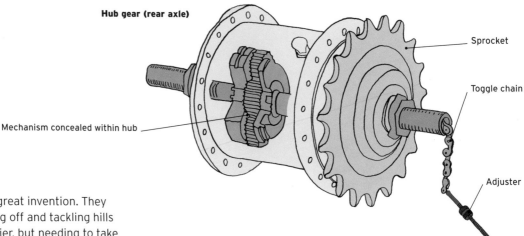

Hub gear (rear axle)

Sprocket

Toggle chain

Adjuster

Mechanism concealed within hub

GEARS

Gears are a great invention. They make starting off and tackling hills so much easier, but needing to take off your shoes and socks in order to count them all doesn't necessarily mean you've got a better bike. Three or five gears may be enough if your habitual route doesn't take you up Mont Blanc. And with a large number of derailleur gears, some may overlap (i.e. be duplicated). Together with the chain and cranks (the arms attached to the pedals), the gears form the transmission system. There are two main types.

Hub

Here the mechanism (or 'mech') is encased in the back wheel hub and is operated by a cable, which is tightened or loosened by the gear shift. The gears are therefore protected from the elements and are reliable, requiring little maintenance, and as a bonus the chain is unlikely to come off mid-ride. You can change gear while stationary, an advantage at a stop sign, for example, if you are in too high a gear for starting off again, and you can change gear while freewheeling (coasting). However, they are heavier and offer fewer gears than derailleurs – usually between three and eight (some premium brands offer more), although for cycling around town that should be enough – and if something does go wrong, hub gears are more difficult to fix.

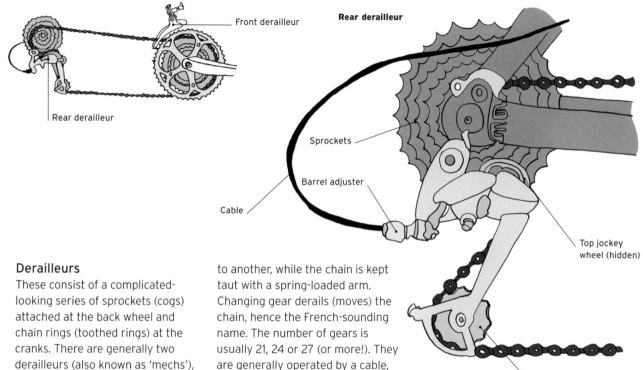

Front derailleur

Rear derailleur

Rear derailleur

Sprockets

Barrel adjuster

Cable

Top jockey wheel (hidden)

Bottom jockey wheel (spring loaded to take up slack in chain)

Derailleurs

These consist of a complicated-looking series of sprockets (cogs) attached at the back wheel and chain rings (toothed rings) at the cranks. There are generally two derailleurs (also known as 'mechs'), one mounted at the cranks (by the pedals), moving the chain from one chain ring to another, and one mounted near the back wheel axle, moving the chain from one sprocket to another, while the chain is kept taut with a spring-loaded arm. Changing gear derails (moves) the chain, hence the French-sounding name. The number of gears is usually 21, 24 or 27 (or more!). They are generally operated by a cable, which is tightened or loosened by the gear shift (on the handlebars, down tube or stem).

Derailleurs need more adjusting than hub gears and as they are exposed they are more vulnerable to dirt and being knocked, but are easier to work on. Unlike hub gears, they are often fitted with quick release rear wheels - better for maintenance and punctures. However, you have to be pedalling to change gear, so you cannot change gear while stationary or freewheeling (coasting).

Gear shift

This is usually a lever, a button on the handlebars, or a twist-grip, where the grips on the handlebars twist around and click into the required gear. On older bikes you may find the gear shift on the down tube. On modern bikes with drop handlebars, the brake and gear shift may be combined in a single lever. Electronic gear shifts are now available for high-end bikes. You press a button or switch and an electronic motor moves the chain across the sprockets. The switch is connected by a wire or wirelessly to a battery pack and the motor.

Good to know
A heavy bike is harder and less responsive when riding on a surfaced road, so the perfect frame strikes a balance between strength and weight, ideally with a nod to flexibility - giving springiness and comfort to your ride - thrown in.

WHEELS

Weight is once again a consideration here and depends on what the wheels are made of. Wheels with narrow rims and tyres (such as on racing bikes) are easier to ride, while wheels with wider rims and tyres (such as on mountain bikes) cope better with potholes but require more effort to ride on surfaced roads. Most wheels have metal spokes (usually stainless steel) radiating out from the hub and poking through the wheel rim where they are secured with a small nut.

The role of the humble spoke is to provide strength and hold the wheel in tension, keeping it 'true' (straight).

Steel wheel rims are durable, but provide a less effective braking surface for rim brakes, particularly when wet.

Aluminium alloy rims are fitted to most bikes these days. They wear more quickly than steel, but are lighter and provide a more effective braking surface for rim brakes.

Good to know
How can you tell if you have steel or aluminium rims? Steel rims (if clean) look shiny, whereas aluminium rims look duller.

TYRES

Tyres grip the road and when correctly inflated provide cushioning to give you a safe and smooth ride. Most road tyres are clincher tyres, where the bead around the inner edges fits inside the wheel rim, and are used with an inner tube. Some high performance bikes have tubular tyres, where the tyres are stitched closed around an inner tube and are glued onto a special rim.

Tubeless tyres are often fitted to mountain bikes. A sealant is generally used to help seal the tyre to the rim - there is no inner tube - and the tyre itself is then inflated.

The tread is the part of the tyre that makes contact with the road. You can change the tyres on your bike to, say, increase or decrease the amount of tread, but tyre sizes are a bit complicated and not standard between manufacturers, so it's best to get your bike shop to advise (see also page 64).

Wide With a thick, knobby tread that continues around the side of the tyre, such as on a mountain bike, these provide grip for off-road cycling. They are stronger and more resistant to punctures, and cope better with potholes, but also make the bike heavier and slower to ride on roads.

Thin With a shallow or even no tread (slick), these roll better than knobby tyres and so make for a lighter ride on the road, but are more susceptible to punctures and cope less well with potholes.
Medium For urban cycling, medium-width, semi-slick tyres - such as often fitted to hybrids - are

TREAD AND GRIP

It may sound counter-intuitive, but bike tyres designed for road use don't need a great deal of tread for grip (traction) - most have a small amount of tread or even just patterning on the centre with a little more around the edges for cornering. Cycle tyres don't aquaplane. Unlike car tyres, the amount of bike tyre in contact with the road is very small, and so the weight of bike and rider pressing down is more concentrated, and is sufficient to have a water dispersing effect. So the greater the tread, the less actual contact with the road. Wider tyres offer more grip, but the quality of the rubber is also important. But no tyre is foolproof, and all can still slip on wet roads, manhole covers, painted lines, etc., so take that little bit of extra care when riding in the wet.

a good compromise, offering the right amount of grip for control without slowing you down too much. Tyres with a serrated surface around the edge are good for gripping the road when cornering.

Puncture-resistant tyres Extratough, these usually have a layer of puncture-resistant material (e.g. Kevlar) and can be slightly heavier than conventional tyres.

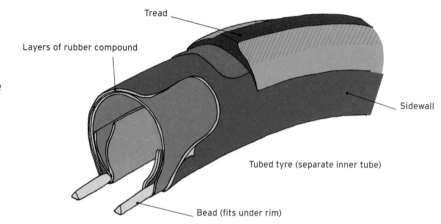

Tread

Layers of rubber compound

Sidewall

Tubed tyre (separate inner tube)

Bead (fits under rim)

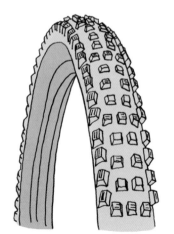

Wide and knobby
(e.g. mountain bike)

Medium and semi-slick
(e.g. hybrid)

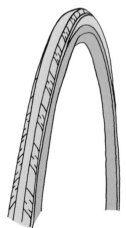

Thin and slick
(e.g. road/racing bike)

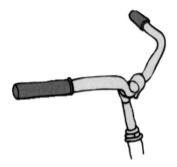

North Road bars sweep up and then down and back towards the rider, are comfortable over both short and long distances and are popular on town bikes.

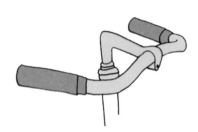

Swept back Curving back towards the rider, these give the classic sit-up-and-beg style that is so great for enabling you to 'see and be seen', but not very aerodynamic for cycling into the wind or downhill.

HANDLEBARS

From bars with armrests for long distance trips, to traditional retro-style handlebars, the choice is wide. Bars that allow you to change the position of your hands help to prevent fatigue and numbness over distance, although for riding in the city this may not be an issue. And generally, the wider the handlebars, the more stable the ride.

Riser rise up to a greater or lesser degree on either side of the stem, allowing the rider to sit a little more upright; they are normally a little wider than flat handlebars, offering more control.

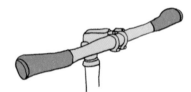

Flat (straight-ish) These offer just one position for your hands, which can be tiring after a while, but they allow for a fairly upright riding position, so you can observe the road well and also be seen.

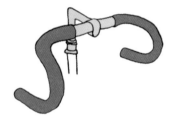

Drop The body leans forward in an aerodynamic position, great for going fast, but less useful for observing the road in traffic. However, you can ride with your hands on top of the handlebars, but your hands usually need to be fairly far forward in order to operate the brakes.

Bullhorn The ends of the handlebars bend up and away from you, useful for stretching out into a more aerodynamic position. Bar ends fitted to flat handlebars produce a similar sort of effect.

Butterfly bars curve around so much that they almost make a complete loop, offering several different hand positions and plenty of scope for attaching gadgets.

BRAKES

There are three basic kinds: rim, hub/drum brakes and disc.

Rim brakes are usually fitted to front and back wheels. When you squeeze the brake lever, it tightens a cable that presses the brake pads against the wheel rim. Rim brakes tend to be less efficient in wet conditions but are lightweight and easy to maintain. There are several main types:

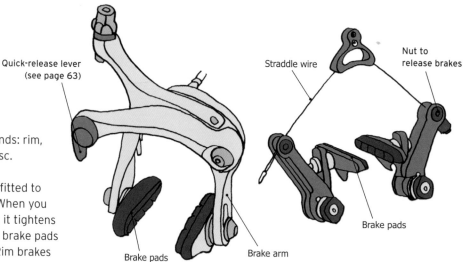

Quick-release lever
(see page 63)

Straddle wire

Nut to
release brakes

Brake pads

Brake pads

Brake arm

Brake pads

Caliper These are fitted to a range of bikes. Two arms squeeze together to press the brake pads onto the rim. They can be single pivot (or side pull, usually on older bikes) or dual pivot (more common).

Cantilever These are less common now. A straddle wire is pulled up to apply the brake pads to the rim. They allow more clearance for wider tyres than calliper brakes.

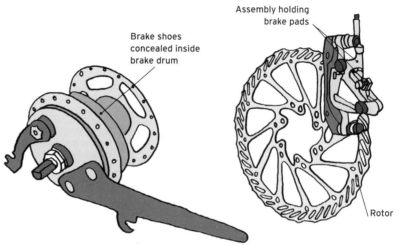

Brake shoes
concealed inside
brake drum

Assembly holding
brake pads

Rotor

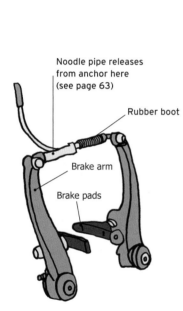

Noodle pipe releases
from anchor here
(see page 63)

Rubber boot

Brake arm

Brake pads

V-brakes Although common on mountain bikes, these are now also fitted to many hybrids and road bikes. They are more powerful than other types of rim brake and offer good clearance for wide tyres.

Hub/drum brakes are protected inside the hub and so require less maintenance. They work by pressing two brake shoes against the inside of the wheel hub, and are operated by a cable or by backpedalling, known as a coaster brake. Coaster brakes are fitted to rear wheels only and can take a little getting used to.

Disc brakes were mostly used on mountain bikes as they work well in muddy conditions, but are now becoming more popular on other bikes. A metal disc (rotor) is attached to the wheel hub, and the brake pads squeeze together against the disc. Disc brakes are powerful and reliable but can be expensive. They can be operated by cable or a hydraulic system.

SUSPENSION

Many urban cyclists are perfectly happy riding a bike with no suspension. You don't really need it around town and certainly not on both front and back wheels, and it only adds to the overall weight of the bike. The best suspension absorbs the shock of riding over bumps and controls downward bounce, so that the wheels follow the contours of the road as evenly as possible. Shock absorbers in the form of a telescopic fork are often fitted at the front to hybrids and mountain bikes. A bike with front suspension only is known as 'hardtail'. The saddle and the seatpost may also have some form of suspension built in and the wheels and tyres (pumped up to the correct pressure) also play their part in absorbing shocks, as does the flexibility of the frame.

PEDALS

There are various different types of pedals.

Open You can move your foot on and off the pedal easily – handy in traffic when you are stop/starting all the time.

Toe clips Racing bikes used to have toe clips or straps to slip your feet into, but these have largely been replaced by clip-in pedals, also confusingly known as clipless, to make the distinction between open pedals and those with toe clips.

Clipless These are used with a special type of shoe, which clips onto the pedal. You press down to engage the clip and twist your foot to release it – the tension/force required to release your foot can normally be adjusted. Although your foot is fixed, there is still a certain degree of movement on the

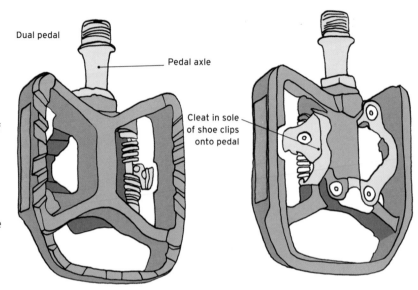

Dual pedal

Pedal axle

Cleat in sole of shoe clips onto pedal

Platform side (normal shoes)

Clipless side (special shoes)

pedal, known as 'float'. Having your foot connected to the pedals makes for better pedalling efficiency as you can pull up the pedal on the up-stroke as well as power down on the down-stroke, and it helps with fancy moves like the bunny hop, but it takes some getting used to. You have to remember to unclip

your foot when coming to a halt or you'll topple over! Dual-purpose pedals are also available – one side is the open, basic platform type for any shoe and the other has a clip. If you intend to walk around in your cycling shoes, make sure you buy those that allow the cleat to be recessed into the soles.

CRANKSET/CHAINSET

The cranks are the arms that connect the pedals to the chain rings, and together they make up the crankset. Cranks come in different lengths, which some believe affects pedalling efficiency and comfort.

SADDLE

Finding a suitable saddle is often a case of trial and error, and you may not know if one really suits you until you have tried it for several hours. The softest, most padded saddle won't necessarily provide the most comfortable ride. Avoid seams on the top that could cause chafing. Some saddles have slots cut into the centre, to help relieve pressure on the perineal region, which can cause numbness. A broader saddle suits a more upright riding position and is more comfortable for women as they have wider pelvic bones than men.

Most bikes today are fitted with lightweight, padded saddles covered with a synthetic cover. The saddle is usually mounted on the seatpost on rails and so can be moved backwards and forwards, and tilted up and down slightly for the most comfortable position.

OTHER FEATURES

Freewheel This feature allows the rear wheel to rotate while the cranks and pedals remain stationary.

Bottom bracket This rotating unit connects the two cranks and pedals to each other. It sits inside the round casing at the very bottom of the frame's down tube.

Transmission/drivetrain If you immediately think 'gears' as in a car, you wouldn't be far wrong. In a bike, the transmission covers the chainset, chain, pedals, sprockets and derailleurs or hub gears.

Headset Connects the frame (head tube) to the forks either side of the the front wheel; contains bearings and allows the wheel to turn.

Stem This connects the handlebars to the forks. It can usually be adjusted slightly for height, but to adjust for reach you need a different length stem.

THE BIKE BRIGADE

There is a huge variety of different bikes on offer. Here is a quick guide to the main types and their features.

Town bike

Fixie

Folding bike

Hybrid

Mountain bike

Town bike Also known as an urban bike, roadster, Dutch bike, etc. A traditional bike with a retro look and an upright riding position. They often come equipped with a basket, rack, mudguard, chainguard and dress guard. Often heavy, they can be hard work up hills. High-quality brands tend to be expensive.

'Fixie' Also lightweight, with just one sprocket on the rear hub so it's also single speed, but no freewheel means there's no rest for the wicked – if you stop pedalling, the bike stops too. The purest of fixies have no brakes (though in the UK, you must have a front brake to be street legal). To stop, the rider eases pedalling, applying resistance, causing the back wheel to slow (and sometimes skid). Not for novices, but devotees mention 'Zen' in the same breath as 'fixie', saying that the bike feels like an extension of their body in a way that a 'normal' bike never can.

Folding bike Great for carrying on public transport (although do try out the folded weight in one hand before buying), but not for cycling long distances. Some models may feel a bit unstable. Bikes of reasonable quality can be expensive.

Hybrid A great all-rounder, with medium-width tyres and an upright riding position, that can be used both for commuting to work and for leisure cycling at weekends. A hybrid is light and fairly easy to ride, with many different models now available at reasonable prices.

Mountain bike A popular choice despite the obvious lack of snow-capped mountains in most cities, with tyres changed to a medium/slick tread for city use. Sturdy but heavy to ride; the suspension helps with potholes. Avoid really cheap mountain bike models: they're unlikely to last.

Fix my fixie

Ultra-hip 'fixies' are the simplest of bikes beloved of cycle couriers and highly skilled city riders, but if you're considering buying one for around town (and you're not a cool courier), you're better off with a single speed (with a freewheel and front and back brakes).

Single-speed bikes Lightweight with one sprocket (single-speed freewheel) on the rear hub, so less maintenance, but can be hard work on hills. The freewheel means you can coast so should have a front and back brake. Some models have a flip-flop rear hub with a sprocket on either side - one side is fixed (as on a fixie), the other a freewheel. To convert from one to the other, you remove the wheel and flip it over.

Touring bike Built to carry plenty of luggage and eat up the miles in comfort, touring bikes are sturdier than racing bikes and usually have drop handlebars, making for a less comfortable town ride.

Road/racing bike These bikes are great for cycling long distances at speed, but the narrow tyres and drop handlebars can make them less comfortable to ride around town.

AND THERE'S MORE

Delving further into areas of specialization, there are even more types of bike, but most are not suitable for urban cycling. Recumbent bikes, where you cycle in a reclining/sitting position, are good for bad backs, but hard on hills and not great in traffic. Press a button on an electric bike and a motor kicks in to help you up a hill. Sounds good? Yes, but the motor adds weight, and you have to charge them up. Tandems are fun but you need a friend. Cargo bikes are designed to carry large goods that can't be crammed into panniers, with an elongated front or back section on which a platform or box is mounted.

SIZING THINGS UP

If you've never bought a bike before it may not have occurred to you that you need to buy the right size, just like a new pair of jeans.

Bikes are available in different sizes but are usually built in proportion, so if you have, say, long legs but a short torso, you may find you have to overstretch to reach the handlebars on a bike that would otherwise seem right for you, so try your prospective bike on for size. Women normally have longer legs and shorter torsos, and men vice versa, so manufacturers design bikes especially for both sexes (see page 28). But don't worry if the perfect bike doesn't seem to exist, as a number of components can be adjusted to fix that.

BACK should be slightly rounded; you should not be sitting up so much that your back curves inwards at the base of the spine.

ARMS (REACH) With your hands resting on the handlebars, your elbows should be bent very slightly. Your arms should not be so stretched out that they are straight with locked elbows, which will cause neck ache and/or back ache.

LEGS Sitting on the saddle, place the ball of your foot on a pedal in its lowest position – your leg should be just slightly bent. If you find your hips are rocking from side to side when you ride, you need to lower the seat a little. If your legs are too bent, pedalling will be uncomfortable and inefficient, and you need to raise the seat a little.

HANDLEBAR HEIGHT You may need to experiment to get this right; for riding around town in an upright position, the bars will normally be a bit higher than the saddle.

HANDLEBAR WIDTH should be roughly the same as your shoulders; too narrow and your breathing is constricted.

Brake levers Check you can reach them easily to operate them quickly and safely. They can usually be adjusted to bring them closer to the handlebars. With your hands resting on flat handlebars, you should be able to ride with a couple of fingers curled over the levers, ready to pull them. Some brake levers on drop handlebars are more suited to large hands, so do check.

Saddle Should be level or almost level. If the front tilts down too much, it throws your weight forward onto the handlebars; if it tilts up, you will slide back. Having said that, if it feels uncomfortable, try lowering the front very slightly. The saddle should not sit right on top of the seatpost - received wisdom says that there should be around 7-10cm (2³/₄-4in) of seatpost visible between the saddle and the frame. It's also a good idea to have sufficient clearance between you (and your crotch) and the top tube when off the saddle and standing straddling the bike.

ONLINE OR IN PERSON?

Like just about everything these days, you can buy bikes online, but nothing can replace the expertise of a good bike shop where staff can listen to your requirements and make recommendations. When you're buying a bike, staff will often allow you to swap certain components like the saddle; it's cheaper than buying another later. New bike models tend to appear in the shops in the autumn, so it's a good time to look out for last year's models discounted - there is usually little difference between the two.

LIGHTS

Using lights when cycling around town at night is more about making yourself visible to others (especially motorists) than lighting up the road in front of you, as street lamps normally take care of the latter. It's also important to comply with the law – otherwise you could be deemed negligent if involved in an accident.

There's a whole fairground attraction of (mostly) LED possibilities out there, and ways of fixing them to your bike, usually with quick-release mountings so you can take them with you when you park. Most lights can be set to emit a steady beam or blink, and some are very powerful – but you don't need a search light in town and you don't want to be accused of trapping some hapless pedestrian in your tractor beam. Look for those designed to emit some light at the sides too, to maximize your visibility. Flashing lights get you noticed, but aren't so good at lighting the way.

You can always light up yourself, too, with extra lights worn on your clothing. A head torch mounted on a helmet for the gold-miner look is practical, as the beam is directed wherever you are looking, and it's also handy for stopping to look at a map...or for a nugget of gold. But these lights should only be viewed as supplementary, not as

Good to know

Dynamo lights are powered by the bike (at least that's cheap) as you cycle along, so as soon as you stop they dim and go out. Some have a stand-light that stores some energy and comes on when you stop. Dynamos are mounted permanently on a bike and so are vulnerable to theft.

replacements for the lights on your bike. And if you are carrying something on a rack, be careful not to cover the rear light by mistake.

In the UK, the law states that you must display lights when cycling 'on a public road between sunset and sunrise'. It's not a bad idea to put them on as soon as the light starts to fade and in low daylight conditions too, such as a murky winter's day. The following is the minimum requirement for lights in the UK, but don't forget that there's nothing stopping you from adding more - be safe, be seen!

Front One white light (if it's a steady beam, it must conform to the BS6102/3 or EC equivalent safety standard), that must be positioned centrally or offside, up to 1500mm from the ground, visible from the front. You must not display a red light at the front.

Pedals For bikes made after 1st October 1985, two amber reflectors (BS6102/2 or EC equivalent) per pedal, one plainly visible from the front and one from the back. Clipless pedals can make complying with this law tricky.

Rear One red light (if a steady beam, it must conform to BS3648 or BS6102/3, or EC equivalent), positioned centrally or offside, between 350 and 1500mm from the ground, visible from behind. One red reflector (BS6102/2 or equivalent), positioned centrally or offside, between 250 and 900mm from the ground, visible from behind. If you have a trailer, you must also have a rear light and triangular reflector with an ECE mark III or IIIA on that.

Flashing lights If your front or rear light is only capable of emitting a flashing light, it must be a minimum of 4 candelas.

BELLS AND WHISTLES

In the UK, bikes at the point of sale have to be fitted with a bell, but once out of the store, there is no law dictating that it needs to be fitted or used on the road.

The law aside, a bell is useful for making your presence known to pedestrians (less so to other road users) and especially around blind corners. There is a wide range of bells and klaxons, from a pleasing ping to a honking horn, but a polite 'Hello!' or 'Coming through' can also work a treat. Unexpected loud sounds can make people stop in their tracks - and yours - which rather defeats the object.

KEEPING CLEAN

In bad weather, you don't just need protection from what's coming down from the skies, you also need shielding from what the road throws up at you. Mudguards on front and back wheels will stop the worst of the spattering from the road. For purists who don't like to spoil the line of their bike but don't want a wet backside, try an emergency mudguard - it slots onto the rails under the saddle and can be folded away beneath it when not required. Chain-guards protect clothes from snagging and an oily print of a cog appearing on your best trousers or leg - they can be partial or full.

Or you can fit a dress guard to the back wheel - it doesn't just protect floaty summer dresses, but stops any trailing coats or luggage carried on a rack from snagging and catching in the wheel.

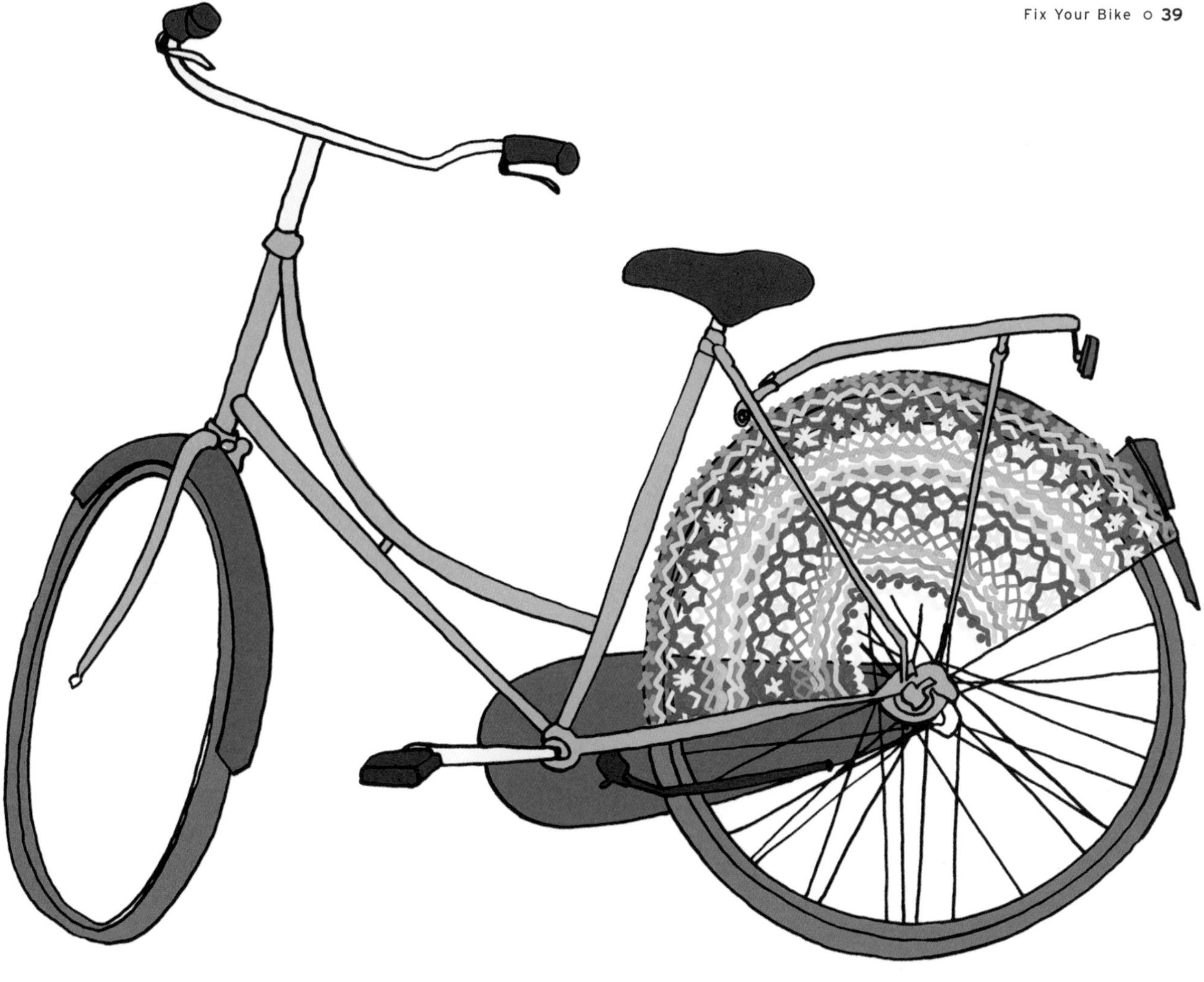

CHAPTER TWO
FIX AND MAINTAIN
SPICK, SPAN AND SPINNING ALONG

BIKE CHECK

How often you check your bike over depends on how often you use it, for how long and where. If you cycle every day in urban conditions, it's good to do a fairly thorough check every 4–6 weeks. The lists on page 42 summarize what to look for and the sections that follow explain in more detail. If you suspect something needs attention, refer to the cross-referenced page for a fix/advice, or get professional help. Don't hesitate to get your bike checked by an expert if something gives you cause for concern as your safety may be at stake.

CHECKLISTS

Here's what you need to be checking for. Pages 44-47 cover each item in more detail.

QUICK PRE-RIDE

☐ Brakes work and release correctly. Squeeze both brake levers hard and try to push the bike forwards - it shouldn't budge.

☐ Disc brakes: check rotors for dirt and fluid.

☐ Hydraulic brakes: check for fluid leaks.

☐ Chain is lubricated and clean.

☐ Quick-release levers/axle nuts on the wheels are secure.

☐ Tyres are pumped up sufficiently, with little or no give when the centre (not the sides) is pressed with the thumb.

PLUS, EVERY 4-6 WEEKS

- ☐ Rim brakes pads for wear and correct alignment, brake cables, brake lever travel.
- ☐ Disc brakes rotors for damage/wobble.
- ☐ Hydraulic brakes hose for wear.
- ☐ Chain for slackness/'stretch'.
- ☐ Chain rings and sprockets for chipped, bent or worn teeth.
- ☐ Front suspension for oil leaks and wear.
- ☐ Gears for slipping/failing to engage.
- ☐ Headset for play and wear.
- ☐ Saddle is secure, won't drop or tilt up down suddenly.
- ☐ Tyres for wear and stones/nails etc. embedded in the tread.
- ☐ Wheels for wobble and play.
- ☐ Wheel rims for wear.

PLUS, KEEP AN EYE ON

- ☐ Cranks/bottom bracket for play and wear.
- ☐ Frame for cracks, bent tubes and other damage.
- ☐ Nuts and bolts for tightness (must be secure, but don't overtighten, which could mess up the thread).
- ☐ Pedals for play and wear.
- ☐ Wheels for loose, bent or broken spokes.

☑ BRAKES

With front and rear brake levers squeezed firmly, try to push the bike backwards and forwards. If the wheels rotate at all, the brakes need attention.

Cable-operated Check the brake cables for fraying, wear or kinks.

Hydraulic-operated Check for fluid leaks and the hose for damage/chafing where clamped to the frame. Check brake lever travel – if it pulls all the way to the bar or feels spongy there may be air in the system, requiring urgent attention.

Rim brakes Check the brake pads for dirt, wear (see page 68) and alignment (see page 70). Check brake lever travel – if the levers touch or nearly touch the handlebars when the brakes are applied, the cable may need tightening (see page 70) or the brake pads replacing (see page 68). Check the brakes are not binding (not releasing when you release the levers). If they are, try lubricating the pivot points and possibly cables (see page 70).

Disc brakes Check the rotors for damage and dirt or fluids (which can cause brake failure). Grasp the rotors and try to move them – there should be no play. Spin both wheels and watch the rotors from above; check they don't rub the brake pads.

☑ CHAIN

Check for 'stretch'/wear using a simple chain-checking tool (follow the instructions). A worn chain (on which the teeth curve over in the shape of a wave rather than stand upright as points), causes wear in the chain rings and sprockets, too: the sprockets are more expensive to replace, but by the time you've noticed the chain is worn (such as if it starts to skip), you'll probably need to replace both. To reduce wear, clean and lubricate the chain regularly, so make this part of your regular maintenance routine (see page 49). To replace a chain that has fallen off, see page 51.

Bike shops
Research a local bike shop you can trust before you need one. Look for a place that offers advice, servicing and repairs. When you take your bike in, ask for an estimate in advance and assurance that if the work is going to cost more, they will check with you first. When you get the bike back, check it right away; if you allow too much time to elapse, it's harder to complain.

Good to know
If you are unlucky enough to be involved in a crash but your bike seems to have escaped unscathed, it's still best to get it checked out at a bike shop (see page 80 for advice on accidents).

☑ CRANKS/BOTTOM BRACKET

Holding the seat/down tube with one hand, try and push each crank in and out - there should not be any play. Listen for clicking or grinding sounds indicating trouble with the bearings. Turn the cranks until the pedals are level on both sides and press down on them together, turn them another 180 degrees and press down again - if the cranks move down a little or you hear a click they need attention and the bottom bracket may need replacing.

☑ GEARS

Run through the gears to check they're not slipping/missing. Using a bike stand makes this easier, or test them on a short ride around the block. You can make minor adjustments yourself (see page 71).

☑ HEADSET

Grip the front wheel firmly between your legs and try to move the handlebars from side to side as well as up and down. They shouldn't move independently of the wheel. Lift the front wheel up off the ground and turn the handlebars, the front wheel should turn freely with them.

☑ PEDALS

Spin the pedals - if they waggle up and down the axle is bent and the pedal will need replacing. Push them in and out on their axles - there should be very little play or movement, if any. If they click, get them checked out.

Good to know

Stop and check your tyres whenever you think you have just ridden over some broken glass. Spin each wheel carefully through a gloved hand to knock out any shards that may be lurking.

☑ TYRES

There are several reasons to keep tyres pumped up to the correct pressure: for a smooth and easy ride, to prolong the life of the tyre, and to help keep punctures at bay. The pressure is marked on the side in PSI (pounds per square inch) or BAR (see page 47). If there's a range marked and you are quite heavy, go with the upper limit, but if that proves a bit too hard and bumpy, take a little air out. To boost grip, such as in wet conditions, go for the lower end - lower pressure means the tyre is flatter and so more of it makes contact with the road, which also means the bike is a bit harder/slower to ride. Don't use a petrol station airline - cycle tyres require far less air than a car and it is easy to over-inflate and blow the tyre up, literally. For advice on using a hand or floor pump see page 52.

Carefully prise out any shards of glass, stones or other foreign bodies. Check the tread for wear and the sidewalls for fraying, bulges and cracks. Numerous small, shallow horizontal cracks are due to normal ageing and are not too serious, but if you want a second opinion as to when to renew a tyre, ask at a trusted bike shop.

☑ WHEELS

With the bike upside down, grasp each wheel and rock it from side to side. A tiny amount of play is fine, but anything excessive needs attention. Spin the wheels – they should spin freely, with little noise and no wobble. With rim brakes, look at the gap between the brake pads and the rim as the wheel spins – it should remain the same, or very nearly. Any variance of more than 2mm (1/12 in) means the wheel needs truing (straightening). Check for any loose or bent spokes. Open and close the quick release skewers (if fitted) to check the wheels are secure in the dropouts, you should need to use a fair amount of force to close them (see page 64), or check the tightness of the axle nuts with a spanner.

Good to know
Tyre sizes (as well as pressure) are marked on the sidewall and can be confusing. Many have an ISO (International Standards Organisation) marking – two numbers separated by a dash – as well as sizes in inches or millimetres. Go by the ISO number or take the old tyre in with you. NB Some tyres are directional (must be fitted so the arrow on the sidewall points in the direction of travel).

☑ WHEEL RIMS

Rim brakes wear out aluminium wheel rims and make them weaker, so check and replace when worn. (Steel rims are less common and much more durable.) Check the built-in wear indicators, e.g. a groove or a tiny indented dot. Feel for a dip in the middle of the rim with your thumb, or press in – does the rim give under fairly strong pressure? Or place a straight edge across the rim – can you see daylight in the middle?

BIKE CARE

Prevention rather than cure applies to more than just medical matters. In terms of bikes, it is far better, and usually cheaper, to be proactive and check and maintain your bike regularly rather than to let circumstances dictate when work has to be done, particularly if those 'circumstances' occur when you're out on the road.

You can clean and lubricate your bike yourself, and do some small maintenance jobs, but take it to a bike shop for advice on anything that concerns you (including the small stuff) and regular servicing.

Good to know

It's a good idea to get your bike serviced at a bike shop regularly – unless you've grown up tinkering with bikes. But even if you're good with bikes, a service at a trusted shop can be time-efficient and cost-effective (even the pros have their own mechanics). Pre-summer and pre-winter are good times, or pre-winter if just once a year.

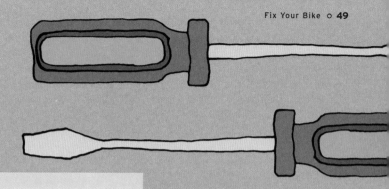

HOME TOOLKIT

No need for a hard hat and reinforced toecaps, a modest amount of tools should see you right.
• Small cross-head and flat screwdrivers
• Star screwdriver (for six-point star screwheads, if fitted on your bike)
• Small spanners
• Adjustable spanner
• Allen (hex) keys
• Tyre levers
• Track pump – or make do with just a hand pump and a pressure gauge
• Rags
• Disposable polythene gloves/hand cleaner
• Bike lubricant

ROAD BIKE KIT

This one is modest enough to tuck into a small bag.
• Allen key multi-tool
• Spanner multi-tool (one suitable for your wheel nuts, if fitted) or adjustable spanner
• Inner tube
• Puncture kit
• Tyre levers
• Frame pump, mini pump, or CO_2 cartridge
• Chainlink removal tool
• Quick/Master link
• Rag
• Disposable polythene gloves/hand cleaner

CHAIN FALLEN OFF?

Replacing a chain that has fallen off is normally an easy fix but can be a messy one, so now's the time to use those polythene gloves in your bike kit.

If the chain falls off mid-ride, stop pedalling immediately to minimise it jamming. If the chain has jammed you will need to prise it free carefully – you'll need something like a screwdriver to do this. (If your chain makes an irritating habit of falling off, add one to your road bike kit.)

Prop the bike against a wall or post if possible, with the sprockets facing you. Lay the chain along the top of the sprockets - you don't need to wrap it right round. To do this, you might need to push the rear derailleur forwards slightly to slacken the tension in the chain. With the chain placed on the front and back sprockets, lift up the rear of the bike with one hand and turn the pedals forwards slowly with the other. The chain should feed itself through and re-engage.

If the chain keeps falling off mid-ride, avoid using the extreme positions of the gear shifts (high or low) until you have had a chance to get it checked out; something could be out of alignment.

Chain reaction
Carrying a chain link removal tool and chain connector (Quick/Master link) will enable to you fix a broken chain en route. Use the tool to push out a pin from a chain link, and the connector snaps into place to repair the break.

PUMP UP TYRES

This number one bike maintenance task is so obvious you may think it barely counts as such, but it's important to ride with tyres at the correct pressure (see page 46).

Pumps

The old-fashioned frame pump clips neatly onto the frame, but is cumbersome to carry in a backpack. Mini pumps are more portable, but hard work and less efficient, though some are double action, pumping air on both the up- and down-strokes. Track/floor pumps are best, being fast and efficient though too big to carry around. Buy one with a built-in pressure gauge. They stand on the floor and you pump with both hands. Be sure to get one that fits your valve, though many have a dual head that fits both types.

Depending on the type of pump, it either connects to the valve directly, or by a hose, and either screws or presses on (there might be a lever to keep it in place). Make sure it is attached to the valve square on in order not to let air out or bend the valve. Don't use a petrol station airline (see page 46).

Valve spotter

Presta valves (usually on tyres inflated to high pressure, e.g. on racing/road bikes) come in different lengths to accommodate different rim depths. Unscrew and remove the cap if fitted, then unscrew the small nut on the top as far as it will go (it shouldn't come off) and attach the pump. Make sure it is on squarely, press down and use the thumb lock if there is one – otherwise, hold the pump onto the valve with one hand and pump with the other. If air doesn't appear to be going in, tap the small nut lightly and try attaching the pump again. To let air out, press down on the tip.

Schrader valves (usually on tyres inflated to a lower pressure, e.g. on mountain bikes) are the same as on a car, so you can also use a car foot pump. Unscrew and remove the cap and press down and attach the pump, using any locking lever. To

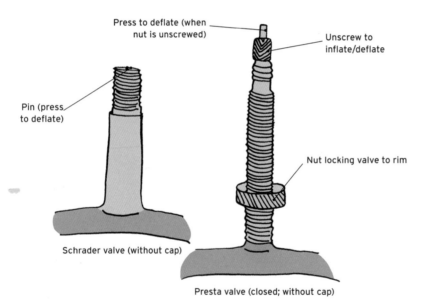

Press to deflate (when
nut is unscrewed)

Unscrew to
inflate/deflate

Pin (press
to deflate)

Nut locking valve to rim

Schrader valve (without cap)

Presta valve (closed; without cap)

let air out, press the central pin inwards.

Woods/Dunlop valve This third type of valve is sometimes found on older bikes. A cross between the Presta and Schrader, it can be inflated with a Presta pump.

Inner tubes

When buying a replacement, always check it's the right size for your tyre. If you're not sure, make a note of the markings on the tyre and ask at the bike shop, or take along the old tube. Take a note of the valve type on the old tube, too – if you buy one with a different valve you may not have the correct pump to inflate it and it may not fit the hole in the wheel rim correctly.

Quick fix
CO_2 cartridges are a quick and simple way to inflate a tyre and get you home. Watch you don't over-inflate; ideally use one that delivers a measured amount that is right for your tyre. It's a temporary fix only: you'll need to reinflate the tyre with a pump within a few hours as CO_2 deflates quickly.

Good to know
Always carry an inner tube: it's usually easier to fit a new tube rather than repair one on the go (at the roadside, in the cold). The repaired one (at home, in the warm) then becomes the spare. The puncture kit is in case you are unlucky enough to get a second or even third puncture.

BIKE WASH

It's surprising how mucky a bike can get, even just around town, so here are some tips on keeping yours spick, span and spinning along nicely. It's not just the cycling equivalent of wearing clean underwear in case you get run over – dirt in moving parts means wear, trouble and expense. In the city it's a good idea to clean little and often (especially in winter, when roads are gritted), with a deep clean every 2–3 months. Using a high-pressure hose won't do your bike any favours, so they're best avoided.

Wash down the bike with hot water and a specialist bike-cleaning liquid, using sponges and brushes to get into all the nooks and crannies, then rinse with clean water. Work from the top down to avoid washing dirt into already clean parts. A brush is good for tackling wheels and spokes, while wheel rims and rotors can be cleaned with a nylon kitchen scourer. Don't forget the brake pads, which may damage rims or rotors by dragging dirt around them – rub them lightly with sandpaper (to release brakes, see page 63). Avoid getting oil onto

CLEANING KIT

Bucket • specialist bike-cleaning liquid • Clean water for rinsing • Degreaser • Rags (cotton is best) • Several sponges (for very dirty bits/less dirty bits/rinsing) • Old toothbrush • Larger brush (e.g. washing-up brush) or specialist cassette brush/scraper • Lubricant (see page 56) • Nylon kitchen scourer • Fine sandpaper • A bike stand is useful if you have one.

brake pads, wheel rims or rotors (disc brakes) – watch out for drips or stray spray and clean straight away with hot water or a degreaser if necessary. Scrub the chain, front and back sprockets, chain rings and derailleurs using a toothbrush dipped in degreaser. To clean between the sprockets, hold a cloth taut and work the edge down between them (or use a cassette scraper). Rinse with clean hot water. If any grease remains, repeat the process. If you have a chain-cleaning machine, follow the instructions. Clean the inside of the forks by passing a rag behind them, pulling it taut and running it to and fro.

Dry the bike off with a clean rag. To dry the chain, hold a rag around it with one hand, turn the pedal backwards with the other and run the chain through the rag to dry it.

That's the spick and span bit, but to keep your bike spinning you now need to lubricate the moving parts.

Good to know
Don't use special cleaning fluids or solvents on wheel rims or disc brake rotors and pads, as they will leave a residue and compromise braking efficiency. Use a specialist disc brake cleaner or hot water and a nylon scourer for the stubborn bits, and rinse thoroughly afterwards.

BIKE LUBE

How often you oil your bike depends on where, when (the time of year) and how often you ride it.

Use lubricant regularly but sparingly, and in the right places – it keeps things running smoothly and helps prevent rust, but it also attracts dirt, which increases wear between moving parts. Old-fashioned oil in particular attracts dirt and is less resistant to water, but is cheaper than synthetic lubricant. The latter is cleaner and comes in 'dry' and 'wet' forms. Dry lubricant is lighter and so is cleaner than wet, but doesn't last so long. It tends to be used more in summer. Wet is heavier and clings more. Clean off old lubricant before using a different type as they may be incompatible and mix up into a mess. It's also best not to keep adding fresh lubricant on top of old, but to get into the habit of first cleaning and then lubricating your bike regularly.

LUBRICATE REGULARLY

✔ **Chain** If it's really dirty, clean it first using a degreaser; otherwise clean it by running it through a rag (see page 54). Lubricate along a free section of the chain (between the chain rings and sprockets), while turning the pedal backwards until the whole chain has been oiled. Then work through all the gears to transfer some lubricant onto the sprockets and chain ring. Now remove any excess by running the chain through a rag again, but holding the rag lightly. If you're using dry lube, the chain should just feel slightly tacky to the touch.

✔ **Pivot points on front and back derailleurs**

✔ **Jockey wheel/rollers**

✔ **Pivot points on brakes** (protect the pads and wheel rims) and brake levers (on drop handlebars, peel back the lever hoods to access the pivot points). Squeeze the levers several times to work the lubricant in.

✔ **Exposed uncoated cables** where they enter the cable housing. (Cables on modern bikes are often coated in a substance like Teflon and don't need lubricating.

LUBRICATE LESS FREQUENTLY

✔ **Pedal axles**

✔ **Bolts**

LUBE KIT
Lubricant with a tube/nozzle attachment for accurate application • Rags • Newspaper or and old sheet to stand the bike on and catch any drips

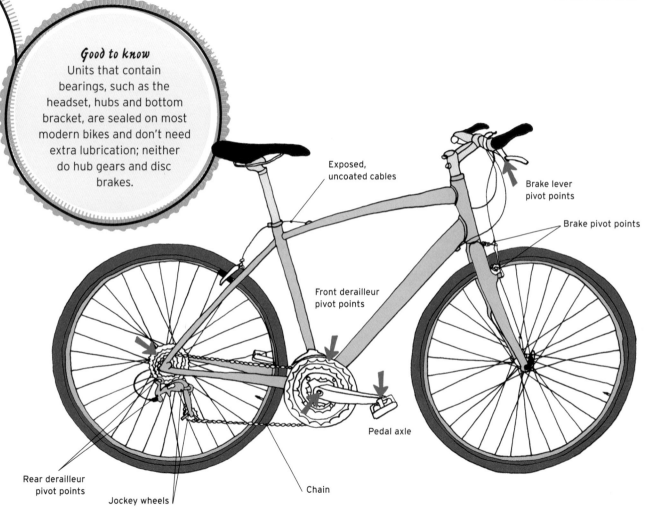

Good to know
Units that contain bearings, such as the headset, hubs and bottom bracket, are sealed on most modern bikes and don't need extra lubrication; neither do hub gears and disc brakes.

Exposed, uncoated cables

Brake lever pivot points

Brake pivot points

Front derailleur pivot points

Pedal axle

Rear derailleur pivot points

Jockey wheels

Chain

BIKE DIY

There are several bike fixes that you can tackle yourself. With your toolkit, the right techniques and a bit of practice, you can save yourself a lot of time and money.

PUNCTURES

Sooner or later, you will be confronted by one of these - the granddaddy of all bike DIY. First, for a quick fix if you are lucky, see page 62. Otherwise, it's time to remove the wheel and find out what huge shard of glass - in reality, usually something tiny and innocuous - has done the damage (1).

Remove the wheel (see page 64). Deflate the inner tube (see page 53) and undo the nut locking the valve onto the rim. To get the inner tube out, you need to extract one side of the tyre from the wheel rim. Pinch the sides of the tyre together with your hands and work it backwards and forwards inside the rim to loosen the bead (the hard

edge of the tyre that tucks into the rim) around the whole tyre.

On one side of the wheel, opposite the valve, push the tyre sidewall in and insert the round end of a tyre lever under the bead. Try not to catch and tear the inner tube. Keeping the top of the lever inside the tyre, lever the bead up and hook the other end onto a spoke. Now move 10-15cm (4-6in) along the tyre both to the right and left and repeat the process. Unhook one of the outer levers and, keeping it flat against the rim to avoid tearing the inner tube, slide it around the rest of the rim under the tyre beading so that the whole tyre becomes free. You can take the whole tyre off or just one side.

Push the stem of the valve up into the tyre so it clears the hole in the rim and remove the inner tube (2).

Pump a little air back into the tube and listen for any air escaping. If you spot a likely mark or nick, hold the tube up to your face and see if you can feel the air escaping (3), or lick your finger and place it over the suspected hole to see if it

bubbles. If no luck, fill a bucket with water and immerse the inner tube in it and watch for a stream of bubbles. Immerse the valve also to check for leaks around it – if it's faulty you'll need a new inner tube. Once you find the puncture, dry and mark it.

Let out any air and patch the hole, following the instructions on your puncture kit (4) or self-

adhesive patch (5). While waiting for the patch to dry, examine the tyre for what caused the puncture and remove it. Check the whole tyre while you're at it. Run your fingers around inside, but watch out for anything sharp (though that is the object of the exercise!), or use a cloth that will snag on a nail or thorn. Check for any protruding spoke ends and file them down, or

cover them with extra layers of rim tape, the fabric or plastic tape that lines the rim. If the tape has deteriorated, replace it now.

Reassemble the wheel. With one side of the tyre inside the rim, inflate the inner tube a little, insert the valve through the hole in the rim, making sure it's not at an angle, and push the inner tube back inside the tyre. Let most of the air out again. Push the valve up slightly inside the tyre, and work the bead at this point back under the rim. Now work your way around the rest of the tyre, pushing the bead back under the rim. When this becomes tricky, try placing your hands on either side of the part that is still out, and work them towards each other, pushing the bead with your thumbs. If it's too hard using just your hands, push the round end of a tyre lever under the bead and lever it back under the rim, being careful not to snag the inner tube. (Special tyre levers are available that make this a whole lot easier.) Replace the nut locking the valve to the rim. Inflate the tyre and job done! Well, once you've replaced the wheel and reconnected the brake.

Puncture protection

Steer clear of gutters where possible as they tend to collect the kind of stuff you don't want near your wheels – sharp stones, glass, etc. Having tyres inflated to the correct pressure also helps them to fend off punctures. You can also fit puncture-resistant tyres (see page 19).

Good to know
Fixing a puncture
comes with practice, so
finding a self-help bike
maintenance group can
help a lot.

Quick fix

You might be able to fix a puncture without removing the wheel if you know where the puncture is. Deflate the inner tube and, using your tyre levers, unhook a section of the tyre on either side of the puncture, gently pull out the affected part of the inner tube and fix the puncture. Remove whatever caused it and check inside the tyre at this point for anything else sharp before replacing the inner tube and tyre. Or use a sealant (see right).

Tubeless tyres

The sealant used with these often fixes small punctures. If not, remove one side of the tyre (see pages 58–59). Pour out the sealant, dry the area around the hole in the tyre and fix the puncture with your repair kit. Apply some of the sealant all around the exposed bead and pour the rest back into the tyre. Work the tyre back into the rim and re-inflate it. If you're out and about, you can fit an inner tube and repair the tyre later.

RELEASE BRAKES

Most rim brakes on modern bikes have one of the following quick-release mechanisms.

V-brakes Pull back the rubber boot between the two brake arms. Squeeze the arms together with one hand, push down on the anchor point and pull the noodle pipe away with the other. To close them, squeeze the brake arms together and reverse the process, then test the brake.

Cantilever brakes One end of the wire that straddles the wheel, connecting the two brake arms, is held by a small nut. Push the brake arm, holding this end of the wire against the wheel rim. Pull the nut away from the brake arm to unhook the cable and release the brake. To close it again, just reverse the process, then test the brake to make sure it's OK.

Caliper brakes Look for a small quick-release lever on one brake arm. Simply raise it to release the brakes. After you close the lever to close the brakes again, test them to make sure they're OK.

Road bikes Some brakes fitted to road bikes are released by pressing a button on the brake lever. Some reset the next time the brake lever is squeezed, others need the button pressing back to its original position.

Hub brakes These make life a bit more complicated; look for a step-by-step video online or get an expert to help.

REMOVE/REPLACE WHEELS

Essential (almost always) for fixing punctures, the ability to remove a wheel is also useful for transporting a bike inside a car. If you have rim brakes, you'll need to release them first (disc brakes don't need releasing), so they don't hamper the wheel when it is removed. If not, you may be able to get away with letting some air out of the tyre and easing the wheel between the brake pads.

Front wheel

With the brake released (see page 63), turn the bike upside down. The wheels on most modern bikes are fitted with a quick-release skewer (with a nut at one end and a lever at the other) that secures the axle in the dropouts at the ends of the forks. Holding the nut still with one hand, lift up the lever (it should be

quite stiff) and turn it a few times to undo it with the other. Turn the lever until the skewer is sufficiently loose for the wheel to come out – there's no need to undo it completely. If you have axle nuts, undo them with a spanner; you don't have to take them right off, just loosen them enough to get the wheel off. You might need to

Good to know

To remove the wheel to put a bike in a car, there's no need to turn it upside down. Just release the brake and the quick release skewer and pull the front forks free of the wheel.

waggle the wheel a bit to get it out of the dropouts if there's a raised lip, or it may just pull out easily.

To replace the wheel, reposition it in the dropouts and check it is centred between the forks. Holding the quick-release lever out straight with one hand, tighten the nut on the other side of the axle with the other, until it is snug. Check the wheel is still centred and close the lever. It should take a fair amount of effort to do so – if it's too easy, lift the lever and turn it a bit. If you feel you are having to force the lever too much, undo the nut slightly and try again. Try to close the lever so it aligns with one of the forks or is pointing inwards towards the bike. This is to help avoid getting it caught on something and releasing accidentally. Re-attach the brake (see page 63) and test it.

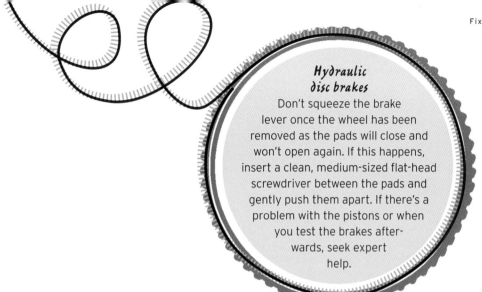

*Hydraulic
disc brakes*
Don't squeeze the brake
lever once the wheel has been
removed as the pads will close and
won't open again. If this happens,
insert a clean, medium-sized flat-head
screwdriver between the pads and
gently push them apart. If there's a
problem with the pistons or when
you test the brakes afterwards, seek expert
help.

Rear wheel (derailleur gears)

This is more tricky due to the chain, and there's no shame in leaving it to an expert if you prefer.

Create some slack in the chain by getting it onto the smallest sprocket at the rear of the bike. If you're not removing the wheel to fix a puncture, just ride the bike to change down. Otherwise, put the bike on a stand, or get someone to hold it up for you. If you don't have help, move the shift down a couple of clicks, hold the rear of the bike up and turn the pedals a couple of times, and repeat until the chain is on the smallest sprocket.

Release the brake (see page 63) and undo the quick-release skewer or nuts securing the wheel in the dropouts (see page 11). If you have horizontal dropouts (where the axle can be in various positions along the slot), note the position so you can replace the wheel correctly. Stand behind the bike and, to keep the chain out of the way, grasp the rear derailleur (with the chain around it) with one hand and pull it towards you, then pull the wheel out with the other hand. You might need to waggle it a bit.

To replace the wheel, stand at the rear of the bike and pull the rear derailleur towards you with one hand, while guiding the wheel

Removing the back wheel

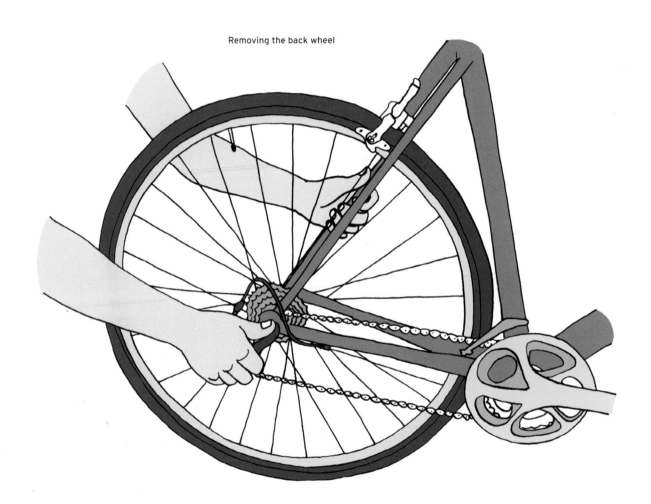

between the forks with the other, so that the smallest sprocket is resting on the bottom part of the chain, then position the wheel in the dropouts. Release the derailleur and make sure the wheel is correctly engaged in the dropouts. Check the wheel is centred before tightening up the skewer or doing up the nuts (see page 64 for further information). Re-attach the brake and test it.

Rear wheel (hub gears)

The procedure varies according to manufacturer, but essentially involves first separating the toggle chain that goes into the hub from the gear change control cable, then unscrewing the nuts on the wheel axle. You can then take the chain off the sprocket and remove the wheel, but the whole process can be fiddly (also refitting it), so look for a video online or use a bike shop.

REPLACE WORN BRAKE PADS

Brake pads need replacing if they are thin (1mm or less of braking material left on disc brake pads), or have worn down to or past any wear indicator line. There are plenty of types of brake pad around, but it's best to replace like with like, particularly if you've never had any problems. It's a good idea to work on one side at a time, so one brake remains mounted as a reference. Once fitted, new pads have to be bedded in, particularly hydraulic disc brakes, so cycle around for a bit on a quiet road or gentle incline and when it's safe, apply the brakes a number of times.

Good to know
When taking things apart, keep nuts, etc. safe in a small container. If you have to remove a succession of nuts and washers, put them down in the order in which they come off, so you know in which order to replace them. Make a note of the way something fits - e.g. curved side down, or take a photo.

Rim brakes

The pads are either removable pads that slide into the existing shoe, or an integrated unit complete with a mounting post, nuts and washers/spacers. Release the brakes (see page 63).

Integrated unit Install and tighten it up lightly. Press in the brake arm with your hand to check the alignment of the pad against the rim. Adjust as necessary and then tighten up the pad fully. If the pad moves out of position as you tighten it, loosen and reposition it, allowing for the skew when tightened. See page 70 for correct alignment and adjusting for toe-in. The washers/spacers are of different thicknesses and can also be swapped from one side of the brake arm to the other to move the pad closer to or away from the rim. Persevere until you get it right – it can be a bit fiddly.

Removable pads Remove any pin or screw holding the pad in the shoe and push it out of the open end. Slide the new pad into position (dunk it in warm water first for a few seconds if it's a bit stiff) and refit the holding pin/screw. Make sure the open end of the shoe points towards the rear of the bike. You shouldn't need to adjust the toe-in (see page 70 for more information), but you may need to adjust the space between the pad and rim (see page 48) to allow for the thicker pads.

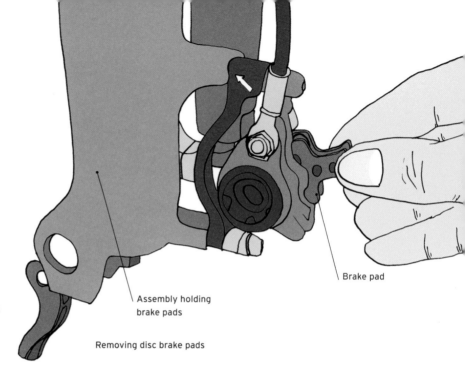

Brake pad

Assembly holding brake pads

Removing disc brake pads

Disc brake pads

Replacing disc brake pads on cable- or hydraulic-operated systems is a straightforward job for a competent bike DIYer, but will vary slightly according to the exact type of brake; check for a video online or with your bike shop. You'll probably need to remove the wheel to change the pads requiring extra tools.

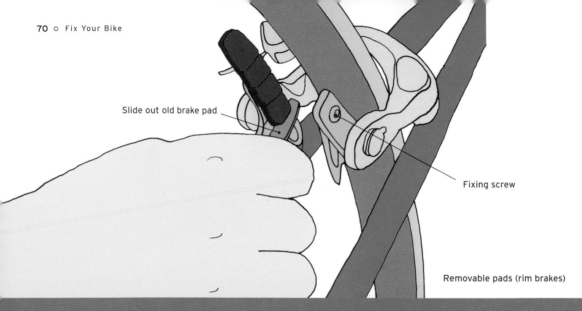

Slide out old brake pad

Fixing screw

Removable pads (rim brakes)

ALIGN RIM BRAKE PADS

The pad must make contact with the centre of the rim squarely, but usually with the front edge making contact with the rim first, just before the rear edge (known as toe-in), to help prevent brake judder/squeal. The amount of toe-in varies, with V-brakes requiring none or very little, whereas some caliper and cantilever brakes should be set so there's a gap of 1mm between the front edge of the pad and the rim and 2mm at the rear. The pads should be about 1-2mm below the top of the rim, not touching the tyre but not hanging down below the rim.

ADJUST BRAKE LEVER TRAVEL

The brakes should begin to apply well before the brake levers reach the handlebars. If you find your levers are touching the bars it may just be a question of replacing brake pads worn thin (see page 68). But otherwise, try tightening the brake cables using the barrel adjuster screw. It is normally where the brake cable enters the brake lever unit, or on the brake itself, where the cable attaches to the brake arm. Undo the lock ring (if fitted) and turn the barrel adjuster a couple of times anticlockwise (usually). On hub brakes, there's an

Good to know
A broken chain can be repaired using a chain link removal tool - just follow the manufacturer's instructions or look for an online video. A quick/master link (in addition) should make it a bit easier, particularly at the roadside - again, follow the instructions that come with it.

extra-long adjuster barrel at the brake arm, which allows more adjustment. Try the brakes and repeat until you are satisfied with the amount of lever travel (and tighten up the lock ring). If this doesn't work or the barrel can't be turned any further, it's a job for the bike shop.

ADJUST GEARS
If some gears slip or won't engage, the cable may have stretched. You can make a minor adjustment yourself that might sort it out, but if it doesn't, it's off to the bike shop.
Derailleur gears Use the barrel adjuster where the cable enters the rear derailleur. If it's hard to shift into a lower gear, turn the adjuster anticlockwise a quarter turn, and vice versa. Make only small quarter

turns at a time as the mechanism is very sensitive, test the gears and repeat until you have sorted the problem. There may also be a second barrel adjuster where the cable enters the gearshift on the handlebars. In which case, you can use either to make the adjustment.
Hub gears Use the adjuster on the toggle chain where it enters the wheel hub and follow the manufacturer's instructions.

CHAPTER THREE
IRIDE
CYCLING SAFELY IN THE CITY

GETTING AROUND

Architects and town planners are thinking bike. They are building cycle lanes and paths, bike-friendly bridges and parks, and designing secure bike stations in towns and cities around the world in more detail. There's still a way to go in some cities but things are on the right track.

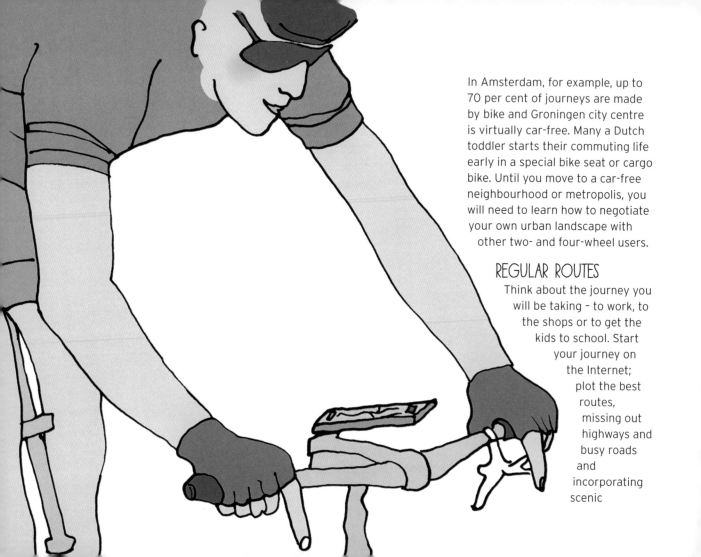

In Amsterdam, for example, up to 70 per cent of journeys are made by bike and Groningen city centre is virtually car-free. Many a Dutch toddler starts their commuting life early in a special bike seat or cargo bike. Until you move to a car-free neighbourhood or metropolis, you will need to learn how to negotiate your own urban landscape with other two- and four-wheel users.

REGULAR ROUTES

Think about the journey you will be taking – to work, to the shops or to get the kids to school. Start your journey on the Internet; plot the best routes, missing out highways and busy roads and incorporating scenic highlights if time and topography allow! Invest in a good city map and study it ahead of time or mount it on your handlebars, where you can also install a satnav or your apped-up smartphone. Watch and learn from seasoned commuters. Check in on local blogs for travel tips and alerts.

Public transport is not bike-friendly in the rush hour due to space issues. On the London Underground, for example, folded bikes can travel anywhere anytime, but unfolding bicycles are not allowed during peak hours (Monday to Friday, 7.30–9.30am and 4.00–7.00pm). Tandems or bicycles with trailers are not allowed, ever.

Folded bikes are accepted on buses at the driver's discretion. Try to be courteous to fellow bike-laden or bike-free passengers. You're all trying to get there as quickly and comfortably as possible.

It is worth doing a trial run of a new journey without the added pressure of reaching your destination at a set time. Do the journey in your planned work clothes to see how it fares with the weather, fair or foul. Here's a checklist of questions to ask yourself if yours is not a folding bike (these can generally be taken on as luggage):
• When can I travel on my local train/bus and when is the best time to avoid crowds?
• Are there designated carriages?
• Do any of my local municipal buses have an exterior bike rack?
• Where and how will I secure my bike and do I need to take extra cords to secure it?

You could always cycle to your local station, park your bike, take the train, and then ride a second, (possibly cheaper) bike locked up at your destination station to work.

LOAD 'EM UP

If you need to transport children or heavy items, various practical and stylish (if pricey) options are on offer for your precious cargo: a trailer that attaches easily with a hitch can carry a serious amount of shopping. Cargo bikes can transport kids (or dogs) and are a practical green option for the school run. They are heavier and slower and parking needs thought as they and their load can attract thieves and vandals, but front-loading cargo bikes present a good alternative to the car. Expensive, electric-assist versions are available, but bike-sharing clubs exist that are worth hooking up to.

TRAVELLING WITH BIKES

When working out a hybrid route (bus/train/tram + bike), check with your national or local train operator on the terms and  (importantly if frustratingly) restrictions on taking your bike on board. There are only a limited number of bike spaces on some trains, so you need to reserve a space before travel (and two spaces for tandems). Always check online for the latest conditions and charges.

Small folding cycles are generally carried free of charge and without restriction, but may need to be in a protective carrying case or stowed as luggage, and rush hour restrictions may apply. Folders can usually be taken as normal luggage on Eurostar (but check first re. maximum size) and European trains. For hassle-free but not fee-free travel, send your bike ahead as registered luggage in a bike bag or box.

UP, UP AND AWAY

Bikes can travel by plane, but check the terms and conditions – and, crucially, luggage allowance – of different carriers first. Some require that you remove the pedals and wheels, let the air out (due to pressure in the hold) and turn the handlebars round, but bagging is also becoming the norm now. Heavy-duty polythene bags, soft bike bags or hard bike cases present different degrees of protection and expense and you have to decide where to leave the bag if you are touring. It helps

USEFUL WEBSITES

www.nationalexpress.com/coach/ourservice/luggage_policy.aspx
Bikes on buses and coaches
www.nationalrail.co.uk/stations_destinations/cyclists.aspx
Travelling with your bike round the UK
www.scotrail.co.uk/cycling Taking bikes on Scottish trains
www.bahn.de (click on English version) for European trains
www.amtrak.com/bring-your-bicycle-onboard What to do in the USA
www.bikeaccess.net Doing it around the world.

when transporting your trusty steed to reduce its height to under a metre by removing the front wheel and putting the saddle down. Pipe lagging helps protect the frame, available from DIY shops.

LEISURE CYCLING

Towpaths were originally the domain of horses pulling working boats along canals, but now offer safe and enjoyable cycling routes for all the family. Find out more on www.sustrans.org.uk. Plan your route carefully and check out the rules of the towpath. Pedestrians have right of way and you should be prepared to give way or dismount if necessary. Ring your bell twice to warn of your approach and pass other towpath users with care. And be respectful of the environment, don't drop litter and try not to fall in the water. Many disused railway routes have been converted into paths and are great for recreational cycling, but they are also there for the enjoyment of walkers and horse riders and can get very busy, so the same rules apply as on other roads. Courtesy and caution in crowds cut the crashes. Care and share.

TOP CYCLING CITIES

The Copenhagenize Index of 2013 ranked 150 cities around the world as pedal paradises, and the top bike-friendly slots featured three Dutch cities: Eindhoven, Utrecht and Amsterdam. Others around the world included Montreal, Munich, Tokyo, Dublin, Berlin, Malmö, Antwerp, Bordeaux and Seville. Barcelona, Paris and Rio de Janeiro get lots of likes for bikes too. New York distributes 375,000 free bike maps showing existing and planned bicycle lanes and routes (www.nycbikemaps.com). Rio has over 450km (280 miles) of cycle lanes and boasts the successful Bike Rio bike-sharing programme; Bogota's CicloRuta network embraces over 300km (186 miles) of cycle lanes and the city closes its streets on Sundays (7am-2pm) and public holidays for the Ciclovia, when cyclists, runners and skaters can enjoy car-free wheeling. So when you have mastered your own environment, head to other cities to explore theirs.

STORAGE

Where to put your bike can be a challenge. Designated bike stations near transport hubs are on the increase in some countries, and workplaces are becoming more bike-friendly. Getting the hang of it at home, where space is often at a premium, can present other difficulties, but solutions abound. Try one of the sturdy plastic multi-coloured bike stands that fix to the wall and onto which you hang the bike(s). Customized wall fittings (of the Nordic variety) help keep bikes aloft in your inner-city loft conversion apartment.

Two bikes can generally cuddle up in an alcove under a flight of stairs, and storage solutions for multi-bike flatsharers or families are on the up. Pulley systems lift them off the floor and up to the ceiling. Think vertically as well as laterally. A shared hall, covered balcony or a not-much-used room (a shower room, for example) are other options. The bathtub is a last but useful resort for those who live alone or generally shower. Behind the sofa is a possibility, and under the bed can work for folding bikes. Club together with your neighbours and get an outdoor bike stand or rack; try a secure bike locker (pricey and it will give your wheelie bins a run for their money as garden furniture). Lobby the landlord to supply. And think about putting a sock on it - stylish bike covers are available for protecting your room and your bike.

STAYING SAFE

As in life generally, however much care you take and attention you pay, accidents can happen. Here's what to do in challenging circumstances. Stay mindful, literally.

1 Injury If you or others involved in a collision have been hurt, seek medical attention. Head (and back and neck) injuries, however apparently slight, can be dangerous. Check yourself over and check your bike. And watch out for any delayed symptoms with a knock to the head - again, get checked out.

2 Insurance Swap details, get vehicle registrations, check for witnesses, inform the police if you think the law has been broken or if a claim for personal injury or damage is involved. Take photos of damage to all bikes or vehicles involved, and of the accident location. Make a note of the when, where, how and who before you leave the scene. Who had their lights on? Who had right of way? Check the insurance of the other party. Always know the terms and conditions of your own insurance.

3 Fault Do not admit liability as this can be held against you later. Was the road at fault? Take photos and notes of any road damage or hazard so you can alert the local council and possibly claim compensation (see page 86 for potholes).

4 Stay calm Do not carry on cycling. Secure your bike and get a taxi or lift home.

Insurance

Check out the best insurance cover, whether it's part of your health, home or automobile insurance, or as a separate/composite policy that embraces not only bike theft, but also collision damage or personal injury, third party or comprehensive. Many people choose to indemnify themselves through membership of a cycling organisation. The National Cycling Charity (www.ctc.org.uk/insurance) is a good place to start.

Lights, camera, action

You can now mount a camera on the front of a helmet, handlebars or seatpost and record your journey – useful for 'eye-witness' record purposes and 'all the fun of the ride' videos.

Cs AND BE SEEN

While sharing the streets with other drivers and riders, read the signals. Don't engage with aggressive drivers in speed games or macho manoeuvres or competitive weaving. It's a road, not a racetrack or a video game. Ride with courtesy, conviction and confidence. With consideration if you can't manage actual kindness. Stay calm and in control. Be a knight of the road, not a nightmare on it.

FROM RUSTY TO ROAD SAFE

If the last time you got on a bike it had only just lost its stabilizers, then getting back on should be as easy as...riding a bike. It will take some getting used to, but take your courage in both hands and go - ideally on a course to help you cycle in traffic, but at least to the park to practise without the extra intimidation of the traffic before tackling a quiet street and then a busy urban environment.

Here are some back-to-bike basics to get under your belt.

1 Mark a straight line with chalk and practise riding in a straight line, looking over your shoulder from time to time (see illustration on page 84). Improve your balance until the wobble factor is minimal.

2 Try to control your bike with one arm and then practise your signals. Hand signals are important to indicate to other road users what you intend to do (turn, change lanes, slow down/stop), so make sure you can do them without wobbling. Hold your left or right arm straight out to indicate the relevant turn. To slow down or stop, or to indicate you are pulling in to the kerb, many cyclists point slightly out and down.

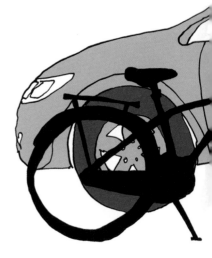

THE A-ZEN OF ROAD RAGE

If things cut up rough on the tarmac, try not to fall victim to road rage (yours or that of others). It is usually a dead-end street. Try to control your temper and be as Zen-like as possible. Respond with humour (or sarcasm if it makes you feel better, and through gritted teeth and clenched jaw, rather than fist, if you must). Or simply issue a non-aggressive 'Hey, didn't you see me, mate? Watch out next time!' A smile and a wave take the heat and the oxygen out of the situation. Hurling abuse, punches or physical objects at drivers or pedestrians, slapping a car bonnet, however much they have done to annoy or endanger you, will get you nowhere (except, in extreme cases, to the police station or hospital ward). Try to practise 'mindful cycling' and to actually enjoy the ride, the scenery, the moment. Cycling is not always just about getting from A to B. Make it about A to Zen.

3 Practise braking. Most bikes have a brake at the front and one at the rear. When you brake, your weight moves forward. The front wheel has more traction and braking power. The rear wheel has less weight and less traction and may lock and skid, so when you need to brake, shift your weight to the rear and increase weight on the pedals while applying the front brake gently (it's the most powerful brake; you don't want to go over the handlebars) and the rear brake, without locking up and skidding.

4 Master the emergency stop. Learn how to come to an abrupt halt; sometimes anticipation is not enough and an adult or child (or pram) will step into the road or a car will come too close. Check your brakes are working correctly and get up to a decent speed, with your hands covering your brakes. Keep your weight to the back of the saddle and apply both brakes, putting slightly more pressure on the front one. If the back wheel feels as if it is lifting, reduce the pressure on the front brake. Practise braking abruptly in wet conditions, when you will need more time and distance. Dry off rim brakes in wet weather and brake a little ahead of steep descents.

TRICKS OF THE TRADE

It is helpful for urban cyclists to master a few techniques to cope with road hazards (and to show off). The **Bunny Hop** is useful for negotiating unexpected potholes and other road obstacles and for pavement mounting in emergencies. In order not to damage your bike on landing, maintenance is important (see page 49) and proper tyre pressure key!

Lift yourself out of the saddle while travelling at a medium speed. Bend arms and knees and press down as if to bounce; pull up the handlebars as hard as you can to lift the front of the bike. To lift the rear, point your toes downwards and 'explode' your legs upwards. This will make the rear end lift off. To avoid a heavy landing, try to level the bike out when you are in the air (it will come with practise). Bend your elbows and knees as you land.

5 If you need to adapt your cycle for disability or difficulties with balance and confidence, there are lots of options out there, including tricycles, tandems, hand cycles and wheelchair cycles. Adapted cycles are the way to go. Literally.

THE KINGS OF HAZARD

These are some of the common cycling pitfalls (literally). Anticipation, looking ahead and good positioning are the three key rules of the road but won't protect you from all incidents and accidents, so keep these warnings in mind:

Gutter rut Don't ride in the gutter. Don't risk being 'doored' by cars. Generally speaking, the space you claim will be the space cars allow you.

Potholey-moly Watch out for these lethal hazards. Don't suddenly swerve out into the road. Always look behind you and signal to drivers. If you just can't avoid

Good to know

Tyred of poor road surfaces? If you find that your route to work is plagued by many of the above hazards, think about fitting fatter tyres, which absorb shock better. Local authorities have a duty to repair potholes and you can report road defects to www.fillthathole.org.uk, or use the app. You can make a claim against the local council for injury or damage caused by a pothole if you can prove that they knew about it – all the more reason to report them in the first place!

hitting one, lift yourself out of your seat, knees bent, and shift your bodyweight to the back of the bike for a softer landing.

Drain pain Bike wheels are exactly the right shape to get stuck in some drainage gates, so watch out for them.

Rain pain Conditions in rain get very slippery, particularly after a prolonged dry spell. Add leaves and mix for the perfect recipe for disaster. Heavy rain can wash debris into the road, and puddles hide potholes, so beware and be aware.

Be on your mettle Metal + water = slippery (manholes, metal drain covers etc). Some raised painted lines can also present similar bumpy rides.

Shard luck If you can't avoid broken glass, stop and check your tyres after negotiating it.

Don't ram the tram If your regular journey involves rail or tram tracks, practise negotiating them. Approach at right angles if you can. If you hit them at an acute angle, you could lose your wheel (wheely painfully – ouch).

Bumpy ride You can usually ride in the space beside or between speed bumps but if a sleeping policeman traverses the whole street, press the bike down as you reach the bump and rock over it. Not if it is the real thing, of course.

VEHICULAR CYCLING: A BRIEF EXPLANATION

John Forester, the 'father of vehicular cycling', coined this term (also known as as bicycle driving) in the 1970s, and its basic tenet is that cyclists 'fare best when they act and are treated as drivers of vehicles'. You need to be confident and fast enough to keep up with traffic to practise this. Some of its advocates oppose segregated bicycling infrastructure, such as protected bike lanes, side paths and specifically designed intersections. In some European countries, cyclists are kept further apart from cars on separate tracks.

LORRYLOADS OF ADVICE

What joy it would be to have the road to yourself. Talking of which, do your very first work or school commute on the weekend or public holiday (or very early morning) to avoid the added pressure of heavy traffic. Cyclists need to be aware of other cyclists and motorbike/car/van/lorry drivers. If you meet a tractor, you may have taken a wrong turning. Be aware of what lorries will and won't be able to eyeball. Police forces now offer the opportunity to sit in a lorry cab so you can witness for yourself what HGV drivers clearly can and clearly can't see from their cabs. If you can't see the HGV's mirrors, the driver can't see

you. In front of a lorry is another blind spot - look behind you. Can you see the driver's head? Then he should be able to see you, but if you can't (his high position may make it impossible for him) move immediately to a safer spot. You have to be the all-seeing eye. Know what's behind and in front and to your side at all times. If you have just passed a lorry, it will still be in the vicinity at the next junction. Know your main junctions. Practise them when commuter traffic is low.
• Don't undertake a lorry or bus at a junction. Overtake or wait behind. Don't cycle in the inside lane.
• If you come up on the inside, as you speed ahead, the lorry could be

turning left, and you won't be making your way dead ahead but the exact opposite. And remember that large lorries often swing out to the right before making a left turn - stay back!
• Don't become sandwiched between guard railings and a vehicle, particularly a large one, at a junction. You could be crushed by a vehicle turning left as there's no room for you to get out of the way.
• Don't cycle on the pavement (where you are not allowed to). You do see cyclists on the pavement in the city, but it is against UK law and

they risk intimidating or colliding with pedestrians, and incurring both wrath and a fine. It is usually a response to tricky road junctions and hostile road conditions where the cyclist simply feels unsafe.

Some pavements have been turned into shared-use paths for bikes and pedestrians. They are not always clearly marked, so if you are on one of these, bear in mind that pedestrians might not be aware that you have right of way, so keep your eyes open. It's not hit and miss – it's a 'don't miss and don't hit' situation.

• Don't be a salmon-biker. Just how silly do fish on a bicycle look? Riding against the traffic (or the wrong way down a one-way street without a bike lane or specific bikes-allowed sign) is not cool, not clever and not funny. Other road users, and pedestrians, will not expect to see you and therefore probably won't a) look for you or b) see you. If folk aren't 'thinking bike', they are even less likely to 'see unexpected bike'. It's against both the law and the law of sense and reason. End of lesson.

• Don't cycle close to a car at speed. Bumper-hugging (or tailgating) is not advised. You don't want to use the back of a car as an emergency stop mechanism. Use your brakes and think ahead instead. Keep enough space between you and vehicles to allow you to do so in time.

• Don't run a red light at a pedestrian crossing. Ever.

• Don't jump a red light at a junction. It's not fine to run red lights. It's a fine (on the spot in many countries).

• Don't make calls or send texts while cycling. You need two hands on the handlebars (and two fingers on the brakes on busy roads) and both eyes and fully engaged brain on the road ahead. Keep your ears open and peeled for other bikes, cars and lorries, although the noise often blends into a general roar in very busy environments. And remember that electric cars and some buses are very hush-hush nowadays. Don't rely on your ears alone – keep looking behind you.

• There's a trend for cycling listening to music on headphones (don't use in-ear phones). After all, what about drivers cocooned in their cars listening to the game or rapping along with Snoop Dogg? But be aware that this reduces your awareness of what's going on around you, and it can also be very distracting. And in the event of an accident, the headphones would almost certainly count against you.

• Don't mix wine and wheels. Cycling under the influence of drink or drugs is against the law. The only bars you really want to spend too long behind are the ones at the front of your bike. You know it makes sense.

• Don't suddenly stop without giving other cyclists and road users warning. Make sure nobody is right behind you and indicate your intention with a raise of the hand, a point to the kerb, or a quick shout ('Stopping!'). Don't cut across cyclists and if you're with fellow riders, let them know what you're doing. Be ready to brake, hands over gears, and then brake gently. When starting off, check all around you, and especially behind, to make sure it's clear. When approaching parked cars from the kerb check behind you, and if safe, pull out a reasonable distance in advance of them; don't lurch out suddenly. Before a manoeuvre, look back and try to make eye contact with the driver behind you, or at least look at the car behind you – hopefully it will make the driver aware of your presence and that you are about to do something. Make clear, decisive hand signals to indicate your planned manoeuvre and execute it. Signalling is as vital for cyclists as it is for drivers. Activate a loud horn or yell to get attention if it looks like the driver is about to pull out without seeing you.

• Give yourself space turning left. Cars are unlikely to do this for you without encouragement. Try to travel towards the centre of your chosen lane as your position A, rather than assuming you have to kerb-cling. You don't have to stick to the designated cycle lane that is part of the ordinary road if it is unsafe to do so. Cars are less likely to collide on turning if they have had to give you room beforehand, and you won't end up in the gutter, as your mother may always have feared. And try whenever possible to put space between you and a parked or even temporarily stationary vehicle to avoid being door-slammed. When is a closed door not a safe door? When it's on a parked car.

• Avoid cycling on very fast roads if you can, unless there is plenty of room for bikes and cars or you can keep up with the pace of the traffic.

In this instance, try to keep as close to the edge of the road as is safely possible so cars can pass. But remember that cars are less likely to see you because they are travelling fast.

• At traffic lights, if you're filtering through traffic, be very aware that drivers may not have noticed you. You may wish to wait your turn, or consider overtaking (on the right) to get to the front, before pulling in to the stream of traffic when it starts moving. Be aware that it's dangerous to undertake on the left (see page 88), and see also cycle lane (page 93). Know how to approach roundabouts and traffic circles. Arrive at and progress in the middle of the most appropriate lane. Take the lane. Cars will be moving more slowly so you can keep up, be more visible to motorists, and avoid being cut up as you exit.

TOP TIPS FOR STAYING SAFE

1 Stay out of the way of bike messengers. They ride all day every day, and it is best to let them get on with their job while you get to yours.

2 Get to know the bike shops on your route to work in case you need air or supplies. They can be helpful and friendly.

3 Give yourself a break. If traffic is very heavy, dismount and use a pedestrian crossing. Better safe than sorry if there's a huge lorry with a blind spot or a roundabout that just looks too treacherous. There's no shame in doing it the safe way.

4 It's hard for a commuting cyclist to go very fast in urban environments, and speeding legislation in the UK covers motorized vehicles only (there is a proposed move to introduce a 15mph (24km/h) limit on some roads on which cyclists would be given priority over motor vehicles). But things are different in royal parks, and sections 28 and 29 of the Road Traffic Act 1988 can be used to report dangerous and careless cycling, so beware.

5 Be aware on designated cycle paths that are separate from the road. This is all fine and dandy when you're on one, but at intersections you're back in the real world where motorists may be unaware of you, and they may cross your lane to get into driveways or entrances. Cycle lanes are very useful but not obligatory. They can also take you into the path of parked vehicles, so be aware of these hazards, too.

CHAPTER FOUR

EQUIPMENT AND SECURITY
KEEPING YOU AND YOUR BIKE PROTECTED

LOCK, STOCK AND KEEP BARRELLING ALONG

There are thieves about, and they love bikes. If you want to thwart them, employ a multi-lock strategy and some wheel cunning. Trees, railings, even traffic poles are no obstacle to those intent on taking your steed.

• Go for a well-lit area and not an isolated back street in which they can work on their lock-smashing or code-breaking undetected.

• If possible, use cycle stands and other designated bike-parking areas bolted or cemented firmly to the ground; avoid racks where only one of the bike wheels is secured.

• In the absence of a rack or solid post, try foiling thieves by locking one bike to another (consenting adults/owners only).

• Try to park next to a flashier model. Some owners camouflage their prize possession with paint or old inner tubes, but this cunning strategy risks the thief thinking that means it is worth taking. In the UK you can register your frame number with the police at bikeregister.com. Write your name and postcode in UV on the frame (invisible to the naked eye) and display the sticker showing that you have done this. Fit an electronic tag for tracking purposes and register at immobilize.com.

• Love it and lock it or lose it, whether your trusty steed is parked on the street, in your yard or even in your house when you are away.

CHOOSING YOUR LOCK

Buy the best lock(s) you can afford. A cheap lock really is a false economy, and the more your bike is worth, the more you should try to invest in making it secure (up to 10 percent of its value, some suggest).

U-lock and D-locks

The U and D locks resemble something the BFG would use as a keyring. Invest in a reputable brand and attach it to both the frame and back wheel, ideally, or the most expensive part of your bike. The shackle lock (a D-lock with a barrel) is another option. Keep the gap between bike and structure to a minimum for thief-thwarting purposes. These locks are not easy to lug around but you can mount them on the bike with a bracket if necessary. Depending on your type of bike you might want to take the saddle with you, or remove the front wheel and lock it behind the bike. They are wheely easy targets.

Chain lock

A heavy-duty chain lock is good for attaching your bike to a bike stand or metal structure. Its strength is determined by the quality of the steel, but it can be vulnerable to covert vandalism by thieves equipped with bolt cutters.

Fix Your Bike o **97**

Cable lock

This might deter opportunists, but use it to attach the front wheel to the frame or in conjunction with a U-lock. A braided cable with lots of thin wire strands is more difficult to cut. If using two types of lock, make them different brands in order to further fox the felon and loop a cable lock through your saddle. Remove the removables: lights, panniers, shopping, and children.

Loop locks or immobilizers

These are generally used on a wheel to prevent your bike from being ridden away. Useful for temporary security during quick stops, they attach directly to the frame but they should not be used for long-term locking. In a bid to find the fool-and-thief-proof lock, design companies have created a bike saddle that comes off easily and transforms into a lock. Other inventions include handlebars that convert into locks. You don't have to carry them around, and in theory the bikes don't get carried away. Result. You can check the security level of various locking devices on www. soldsecure.com.

Document your bike

It's always advisable to keep a photograph and a record of the serial number of your bike, plus the sales receipt. Check if it is covered on your house insurance policy for its full replacement value, or organise separate insurance. And it's wise to keep a spare set of keys for your locks at home and at work, and the code safe. Note the registration number of keys, as some lock retailers replace lost ones.

FIND THE RIGHT GEAR

Cycle chic or lycra-geek? You don't have to dress for your daily commute as if competing for the Maillot Jaune. This chapter will help you gear up for gears.

HELMETS

First of all, the most controversial piece of gear – a helmet. You might think that this is a no-brainer (ouch), but check out some of the lengthy ongoing arguments on the web and you will learn that there is some evidence that helmets encourage drivers to be less careful around cyclists, and in a serious crash, or in a crash at speed, a helmet is unlikely to protect you sufficiently anyway. And do they make cyclists may feel more secure and so take more risks? Does wearing them detract too much from addressing the main cause of cycling accidents being the poor design of roads, bringing bikes and cars into conflict? On the other hand, if you fall off your bike and knock your head on the kerb, you will probably be very glad indeed that you were wearing a helmet. It acts as a shock absorber for the most vulnerable part of the body. Helmets are compulsory in a few countries, including some states in Australia. Ultimately, it's your call.

That said, many cyclists like to wear a helmet, in which case a quality one that fits properly is an absolute must. And it doesn't have to put a spoke in your fashion wheel – there is a huge choice out there, including trendy bucket-style options in a rainbow of colours and styles and with a range of covers, – but don't fall into the shoe trap and buy one that's just a bit tight or a bit loose just because it matches your outfit. On the

practical, rather than fashion, front, vents help you keep cool by allowing air to flow through the helmet. They have removable pads to absorb sweat and provide a cushion, and an optional visor to shade and protect your eyes.

Never buy a second-hand helmet as you don't know what it's been through. It may look fine but the inner foam could be damaged. For the same reason, always replace your helmet if it gets a knock; some manufacturers recommend replacing a helmet every three to five years as the materials can deteriorate, offering reduced protection.

TICK BOX FOR HELMETS

✔ Complies with national safety standard.
✔ Has adjustable buckles.
✔ Is less than 3–5 years old.
✔ Is a bright, visibility-enhancing colour.
✔ Sits level across your forehead.
✔ Fitting pads touch all the way round.
✔ Feels secure and comfortably snug against the chin strap.
✔ The Y-fitting (where the straps come together) is just below your ear.
✔ Hugs your head when you open your mouth really wide.
✔ Brim of the helmet barely visible when you look up.
✔ Fits snugly on your head and is impossible to remove without unbuckling.

SEEING THE LIGHT

Head torches with elastic headbands are available for easy attachment to helmets, with a low setting for road riding with street lights, medium for partially lit roads and full for darkness. Some come complete with remote controls that clip onto your jacket.

THE EYES HAVE IT

Sunglasses protect you from harmful UV rays, wind, rain, grit, dirt, insects and animal muck. Cycling-specific ones should fit helmets snugly and not hinder it in any way (wear the arms over your ear straps, not under) and if budgets allow, consider investing in a pair that adjust with the light or have interchangeable lenses. Check that the frame doesn't obscure your field of vision.

Fit is important, and silicon grippers where the arms contact your head and ears can help keep them in place.

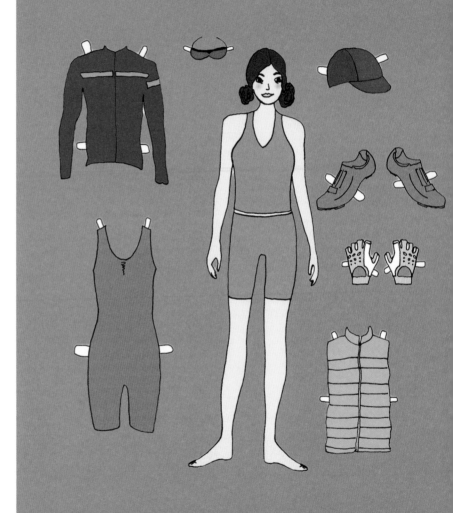

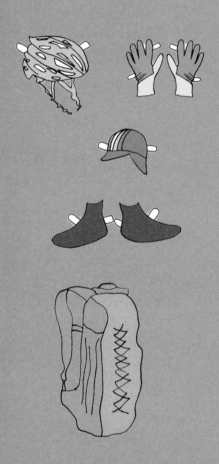

WORKING IT, NOT SKIRTING IT

What you wear will depend on your personal style, your destination and the type and length of your journey. If you are likely to arrive hot and sweaty and can shower or change easily at work, then think about wearing different clothes for the ride, but there is no need to invest in a complete lycra wardrobe unless you really want to. Mammmils (middle aged (mostly male) movers in lycra) are not an endangered species or a universally aesthetically pleasing one. The growth in crossover clothing, cycling gear and urban style is welcome in many ways. Big name designers are incorporating the need for stretch and comfort in their clothes, trousers in particular. There are many more options now for wearing 'destination clothes' on your bike.

Chic and practical commuter jeans, chinos with reinforced seats and quick-drying, water-resistant or full-on rainproof trousers are available for men and women to sport from saddle to desk; functional, with seamless gussets, extra zips, reflective linings and buttons or poppers for the roll-up (trouser leg), they deal with rides and rain in style. For bike-to-boardroom power skirts, look out for smart tailored skirts that unzip at the back to reveal extra fabric to help with movement, and with a high waist for that vital extra coverage. Short skirts by Seymour Leg – that old joke – applies rather specifically to cycling. Long skirts can get caught up in the spokes, so you might want to invest in a dress guard or some clips. You can, of course, wear shorts or leggings under your skirt to protect your legs and your virtue. If you do need to transport a suit to the office (it's a good idea to leave a spare one there), suit carriers are available, some of them actually incorporated into waterproof jackets.

JACKET AND TREWS

A shell jacket is a cycling wardrobe essential to help protect you from the cold, wind and rain. Waterproof and breathable fabric (e.g. Gore-Tex) helps avoid the BIB ('boiling in the bag') feel and look. Look for zipped pit vents to keep you well aired under the arms. Reflective patches and strips are often built-in features of cycling-specific jackets, along with an attachment for a rear light. Some pack neatly away into a pocket and come in hi-vis and fluorescent colours. When choosing waterproof trousers, check for stretchability around the hips and knees, lower-leg zips for putting on/taking off over shoes, and Velcro adjusters to protect legs from the chain. Combine style with safety and check out hyper-reflective clothing that works in the daylight and in the dark. Speed merchants will prefer tight-fitting styles with minimal flapping material, while commuters may prefer looser fits.

iBikepad

Comfort-enhancing, padded shorts under your work gear (for your 'sit bones') deliver a disguised, smoother commute. There are ranges specific to both men and women. Choose something that fits you well. Padding is important, and today it comes complete with breathable and anti-bacterial properties.

LOOK MUM, NO HANDS

Summer cycling gloves and fingerless mitts are good for grip, shock absorption, comfort and palm protection if you fall, and going without gloves is never advisable in winter. Gloves help you maintain a core temperature, as it is your extremities that are the first to lose heat. Invest in some cycling-specific gloves with palm pads and adjustable wristbands or fitted, non-bulky cuffs for draught exclusion. Windproof and waterproof gloves for winter are advisable. Some gloves have a built-in fabric sweat/nose wipe. Remember you need to be able to feel the brakes and gears through the gloves. Men's and women's gloves are different in finger length and wrist size so don't share, borrow or be tempted by the wrong hand-me-downs.

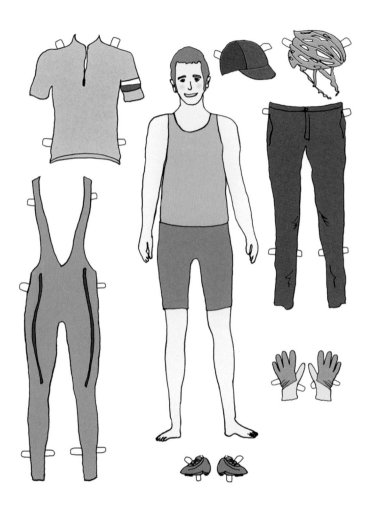

ON THE HOOF

High heels on two wheels don't make for a transport of delight on long, wet or cold journeys. Heels can get stuck in the pedals or in drains and other road hazards. You could so easily make your Manolos into 'Oh nos' (flip-flops are already no-nos). Slip-ons are not ideal, and trainers tend to allow your foot to flex too much, potentially causing damage to legs and hips. If you cycle in your work shoes, go for flats with a stiff sole or, if you have to, wedges. For clipless pedals (see page 26) designers have come up with a range of cycling-specific leather shoes with grippy rubber soles and slight, chic heels that fit neatly on the pedal to give you a safe and stylish ride. Flat walking shoes or cycling-specific footwear are the best options for men. Anti-bacterial merino wool socks keep toes toasty and tasty. And there are always overshoes for when the rains set in.

ALL-WEATHER CHECKLISTS

**Check out these checklists
before you venture out on your
bike in rain, shine or snow.**

RAINDROPS KEEP FALLIN' ON MY BIKE

✔ Cycling cap under a helmet to prevent air-vent drips
✔ Lightweight wet weather jacket that folds and fits into a
  pocket for surprise showers
✔ Rain jacket and waterproof trousers for those cats-and-dogs
  days (check for taped seams, waterproof zips, gaps at your
  neck, wrist, ankle, lower back where the rain can seep in).
✔ Waterproof cape or poncho
✔ Clip-on mudguards (see page 38)
✔ Waterproof socks
✔ Overshoes with zip or Velcro fastening
✔ Waterproof gloves
✔ Bike windscreen when transporting kids on a front seat

YOU ARE THE SUNSHINE OF MY RIDE

✔ High-factor sun cream
✔ Face wipes and deodorant
✔ Clean T-shirt
✔ Cycling cap under a helmet with
  vents to protect your head from
  direct sunlight
✔ Water bottle
✔ Cycling-specific sunglasses
✔ Fingerless gloves or mitts
✔ Regularly washed helmet pads

For advice on lights, bells and mud and
dress guards, see pages 35-39.

IT'S AS COLD AS ICE

✔ Layers are the way to go
✔ A base layer that allows sweat to evaporate. Merino wool keeps you warm, dry and doesn't scratch. You can wear it to work too
✔ An insulating layer, such as a fleece, the weight of which may change
✔ Shell layer that is wind- and waterproof and breathable
✔ Face guard with hi-vis patches and holes for breathing
✔ Neoprene toe covers for the front of your shoe
✔ Arm and leg warmers in bright yellow to keep you warm and seen

CARRY ON BIKING

Carrying your work, gym, going-out clothes and equipment, documents or after-work shopping on a bike presents a few challenges, for which a wide range of practical and chic solutions is available.

PANNIERS

For pannier rack-compatible bikes (check upon purchase), the side, front or back pannier is a practical and stylish answer. Designers have come up with messenger bags and rucksacks that transform into pannier bags, making it easier to transport and convert to a personal bag when you arrive at the workplace. If you are travelling with your laptop, ensure there is sufficient padding in the model you go for, and plenty of secure pockets including an easy-access one for speedy keys, wallet and phone retrieval. Hi-vis covers for rucksack panniers are also available, thereby performing a dual function. If you're heading somewhere overnight or off for the weekend, a combo of two or three panniers (plus rucksack) is a good luggage solution. Brompton folding bikes (and others) have custom-designed bags. A porteur shelf rack sits over the front wheel and becomes a base for a large box bag for bigger loads.

If you're carrying weight regularly, keep an eye on your brake pads and on your rear wheel.

RUCKSACK ON MY BACK

A rucksack is the choice of many urban cyclists. Go for a comfortable design that doesn't affect your ride too much in terms of weight and position on your back. It should be waterproof and equipped with ventilation features to avoid SBS (sweaty back syndrome). Before buying, list what you will be most likely to carry (e.g. check it has a specific laptop and/tablet pouch), and check that your selected bag has enough capacity without being too heavy.

Small but useful

You can keep your tools (and maybe a lipstick or small torch) in an under-the-seat saddlebag. Bar bags can be attached to the handlebars, including those designed to keep your smartphone dry and handy. Seatpost bags attach to the seatpost (not by coincidence) by a tube and they float behind you. Hip packs can work on hips and as over the shoulder holders. They come with tabs for securing to a belt or other bag. Hip.

Keep contents secure in your basket and if you're using bungees with metal hooks, make sure they don't come loose and smack you in the face.

POP IN, POP OUT

The basket can be both a useful and stylish container of equipment or shopping. Back and/or front (hooked onto handlebars or racks), they can be in wicker or metal and covers are available for rainy days. Pop what you need in, head off and take stuff out. What could be simpler? Waterproof baskets with a roof complete with zippered windows and washable floor padding are pet-carrying alternatives for those with furry friends on the hoof.

USEFUL INFORMATION

BOOKS
Richard's 21st Century Bicycle Book by Richard Ballantine
(Pan Books, 2000) the seminal book on cycling in its revised edition
Ultimate Guide to Bicycling Maintenance MagBook
by Guy Andrews (Dennis Publishing, 2010) for those wanting to know
and do more

FILMS
Jour de Fete (Jacques Tati, 1949)
Boy and Bicycle
(Ridley Scott, 1965)
Breaking Away (Peter Yates, 1979)
The Flying Scotsman
(Douglas Mackinnon, 2006)
The Kid with a Bike
(Jean-Pierre and
Luc Dardenne, 2011)

WEBSITES

www.sheldonbrown.com Information from the famous American bicycle mechanic (1944-2008) with encyclopedic knowledge of all things bike
www.tfl.gov.uk/modes/cycling/santander-cycles/community Bike hire in London
www.bikeright.co.uk/services2/cycle_training/adult_cycle_training and **bikeability.dft.gov.uk/the-three-levels/cycling-skills-for-adults** Learn how to cycle in traffic
www.sustrans.org.uk Find out more about the National Cycle Network
www.ctc.org.uk/insurance Get insurance information from the National Cycling Charity
www.tweedride.co.uk and **www.tweedrun.com** For those whose cycling creed on their steed is elegant tweed rather than lycra speed

For travelling with bikes, see page 76.

AND THERE'S MORE.

• Check out with your employer if they have signed up to a scheme whereby you are eligible for a 'cycle to work' scheme that allows you to buy a new bike plus gear at a considerable reduction in price, payable monthly.
• Check out if your city is part of a 'get paid to cycle' program under which commuters are paid to take two wheels instead of four to work.

BIKE-O-PEDIA

The variations in materials and design of every single part of a bike are seemingly endless, from pedals and handlebars to complicated moving parts like derailleur and hub gears. This also applies to the care and maintenance of all the different designs of bike and how their parts fit together and work. We can only cover the most common and give general guidance, but when in doubt there is a wealth of information available online, in books and of course at your local friendly bike shop.

First published in the United Kingdom in 2015 by
Portico
1 Gower Street
London
WC1E 6HD

An imprint of Pavilion Books Company Ltd

ISBN 978-1-91023-202-6

A CIP catalogue record for this book is available from the British Library.

10 9 8 7 6 5 4 3 2 1

Reproduction by Mission Productions Ltd, Hong Kong
Printed and bound by 1010 Printing International Ltd, China

This book can be ordered direct from the publisher at www.pavilionbooks.com